Modern Diesel Technology: Electrical/Electronic Systems and Heating, Ventilation, & Air Conditioning Job Sheets

Dale McPherson

DELMAR
CENGAGE Learning™

Australia • Brazil • Japan • Korea • Mexico • Singapore • Spain • United Kingdom • United States

**Modern Diesel Technology:
Electrical/Electronic Systems and Heating,
Ventilation, & Air Conditioning, Job Sheets**
Dale McPherson

Vice President, Technology and Trades ABU:
  David Garza

Director of Learning Solutions: Sandy Clark

Managing Editor: Larry Main

Senior Acquisitions Editor: David Boelio

Product Manager: Sharon Chambliss

Marketing Director: Deborah S. Yarnell

Marketing Manager: Erin Coffin

Marketing Coordinator: Patti Garrison

Director of Production: Patty Stephan

Production Manager: Andrew Crouth

Content Project Manager: Barbara L. Diaz

Editorial Assistant: Andrea Domkowski

© 2007 Delmar, Cengage Learning

ALL RIGHTS RESERVED. No part of this work covered by the copyright herein may be reproduced, transmitted, stored or used in any form or by any means graphic, electronic, or mechanical, including but not limited to photocopying, recording, scanning, digitizing, taping, Web distribution, information networks, or information storage and retrieval systems, except as permitted under Section 107 or 108 of the 1976 United States Copyright Act, without the prior written permission of the publisher.

> For product information and technology assistance, contact us at
> **Cengage Learning Customer & Sales Support, 1-800-354-9706**
> For permission to use material from this text or product,
> submit all requests online at **www.cengage.com/permissions**
> Further permissions questions can be emailed to
> **permissionrequest@cengage.com**

ISBN-13: 978-1-4180-6338-2

ISBN-10: 1-4180-6338-X

Delmar
Executive Woods
5 Maxwell Drive
Clifton Park, NY 12065
USA

Cengage Learning is a leading provider of customized learning solutions with office locations around the globe, including Singapore, the United Kingdom, Australia, Mexico, Brazil, and Japan. Locate your local office at **www.cengage.com/global**

Cengage Learning products are represented in Canada by Nelson Education, Ltd.

To learn more about Delmar, visit **www.cengage.com/delmar**

Purchase any of our products at your local bookstore or at our preferred online store **www.CengageBrain.com**

Notice to the Reader
Publisher does not warrant or guarantee any of the products described herein or perform any independent analysis in connection with any of the product information contained herein. Publisher does not assume, and expressly disclaims, any obligation to obtain and include information other than that provided to it by the manufacturer. The reader is expressly warned to consider and adopt all safety precautions that might be indicated by the activities described herein and to avoid all potential hazards. By following the instructions contained herein, the reader willingly assumes all risks in connection with such instructions. The publisher makes no representations or warranties of any kind, including but not limited to, the warranties of fitness for particular purpose or merchantability, nor are any such representations implied with respect to the material set forth herein, and the publisher takes no responsibility with respect to such material. The publisher shall not be liable for any special, consequential, or exemplary damages resulting, in whole or part, from the readers' use of, or reliance upon, this material.

Printed in the United States of America
3 4 5 6 7 12 11 10

Contents

Preface vii
NATEF Cross Reference Guides viii

MDT: Job Sheets – Safety 1

Job Sheet 1 - Demonstrate Proper Lifting Procedures 3

Job Sheet 2 - Operate Vehicles Safely in the Shop 5

Job Sheet 3 - Shop Housekeeping Inspection 7

MDT: Job Sheets – Electrical / Electronics 11

Job Sheet 1 - Wiring Diagrams 13

Job Sheet 2 - Checking for Continuity 15

Job Sheet 3 - Checking Current Flow 17

Job Sheet 4 - Checking Resistance 19

Job Sheet 5 - Opens, Shorts, and Grounds 21

Job Sheet 6 - Parasitic Battery Drain Problems 23

Job Sheet 7 - Circuit Protection Devices 25

Job Sheet 8 - Spike Suppression Devices 27

Job Sheet 9 - Frequency and Pulse Width 29

Job Sheet 10 - Battery Load Test 31

Job Sheet 11 - Battery Charging 33

Job Sheet 12 - Jump Starting 35

Job Sheet 13 - Battery Capacitance Testing 37

Job Sheet 14 - Starter Circuit Testing 39

Job Sheet 15 - Starter Control Circuit 41

Job Sheet 16 - Remove and Replace Starter 43

Job Sheet 17 - Charge Indicators 45

Job Sheet 18 - Charging System Diagnosis 47

Job Sheet 19 - Alternator Belt Drive Systems 49

Job Sheet 20 - Charging Circuit Voltage Drop Testing 51

Job Sheet 21 - Leakage of AC Voltage 53

Job Sheet 22 - Headlights 55

Job Sheet 23 - Lighting Components 59

Job Sheet 24 - 7-way Trailer Cord 61

Job Sheet 25 - Using On-board Computer Self Diagnostics 63

MDT: Job Sheets - Heating, Ventilation, & Air Conditioning 65

Job Sheet 1 - Unusual Operating Noises 67

Job Sheet 2 - Unusual Operating Conditions 69

Job Sheet 3 - Performance Testing of Different Types of HVAC Systems 73

Job Sheet 4 - Temperature Control Problems 75

Job Sheet 5 - A/C System Pressure and Temperature Problems 77

Job Sheet 6 - Unusual Operating Noises 81

Job Sheet 7 - A/C System Leak Testing 83

Job Sheet 8 - Evacuate A/C System 85

Job Sheet 9 - Contaminated A/C System Cleaning 87

Job Sheet 10 - Charging A/C System 89

Job Sheet 11 - Identify Lubricant Type Needed for System Application 91

Job Sheet 12 - A/C System Protection Devices 93

Job Sheet 13 - A/C Compressor Belt Drive System 97

Job Sheet 14 - A/C Compressor 99

Job Sheet 15 - Lubricant Level 101

Job Sheet 16 - Lines, Hoses, Fittings and Seals 103

Job Sheet 17 - A/C Condenser Maintenance.. 105

Job Sheet 18 - Receiver/drier and Accumulator/drier Service 107

Job Sheet 19 - Orifice Tube Replacement 109

Job Sheet 20 - Evaporator Housing 111

Job Sheet 21 - Window Fogging Problems.. 113

Job Sheet 22 – Engine Cooling System... 115

Job Sheet 23 - Recover, Flush, and Refill Cooling System 117

Job Sheet 24 - HVAC Electrical Control Systems.. 119

Job Sheet 25 - HVAC Electrical Control System Components 121

Job Sheet 26 - Air, Vacuum and Mechanical Controls... 125

Job Sheet 27 - Refrigerant Recovery, Recycling, and Handling 127

Job Sheet 28 – Handle, Label, Store, Test Recycled Refrig - Non-condensable Gases 129

Preface

The 56 job sheets in this manual have been designed to provide hands-on shop experience to support classroom instruction in two areas of diesel technology. These two areas are Electrical/Electronics and Heating, Ventilation & Air Conditioning. Because these job sheets focus on developing area-specific competencies, they are ideally suited to accompany the corresponding textbooks in the Delmar's Modern Diesel Technology Series. These textbooks are *MDT: Electricity & Electronics* by Joseph A. Bell and *MDT: Heating Ventilation, Air Conditioning & Refrigeration* by John Dixon.

Where applicable, the job sheets are keyed to the task list for Electrical/Electronics and HVAC that underpin NATEF's Medium/Heavy Truck Program Standards. All P-1 tasks in both areas are covered, along with selected P-2 tasks. See the following two pages for a correlation grid of job sheets to specific NATEF tasks. These job sheets provide a means for students to record their progress in mastering diagnostic and service procedures, as well as to capture feedback from instructors. Safety is emphasized solely in the first three job sheets, as well throughout all other job sheets.

Cross Reference Guide

Electrical / Electronics

NATEF Task	Priority	Job Sheet Number
A.1	P-1	1
A.2	P-1	2
A.3	P-1	2
A.4	P-1	3
A.5	P-1	4
A.6	P-1	5
A.7	P-1	6
A.8	P-1	7
A.9	P-3	8
A.10	P-3	9
B.1	P-1	10
B.2	P-1	10
B.3	P-1	10
B.4	P-1	10
B.5	P-1	11
B.6	P-1	10
B.7	P-1	12
B.8	P-2	13
C.1	P-1	14
C.2	P-2	15
C.3	P-2	15
C.4	P-2	16
D.1	P-1	17
D.2	P-1	18
D.3	P-1	19
D.4	P-1	20
D.5	P-1	20
D.6	P-2	19
D.7	P-2	20
D.8	P-1	21
E.1.1	P-1	22
E.1.2	P-1	22
E.1.3	P-1	23
E.1.4	P-1	23
E.1.5	P-2	23
E.1.6	P-2	23
E.1.7	P-1	24
E.2.1	P-1	23
E.2.2	P-1	23
E.2.3	P-2	23
F.1	P-1	25
F.5	P-2	23

Cross Reference Guide

Heating, Ventilation, & Air Conditioning

NATEF Task	Priority	Job Sheet Number
A.1	P-1	1
A.2	P-1	2
A.3	P-1	3
B.1.1	P-1	4
B.1.3	P-1	5
B.1.4	P-1	6
B.1.5	P-1	7
B.1.6	P-1	8
B.1.7	P-2	9
B.1.8	P-1	10
B.1.9	P-1	11
B.2.1	P-1	12
B.2.3	P-1	13
B.2.6	P-2	14
B.2.7	P-2	14
B.3.1	P-1	15
B 3.2	P-1	16
B.3.3	P-1	17
B.3.4	P-2	17
B.3.5	P-1	18
B.3.7	P-1	19
B.3.9	P-1	20
B.3.10	P-1	5
C.1	P-1	4
C.2	P-2	21
C.3	P-1	22
C.4	P-1	22
C.5	P-1	22
C.8	P-1	23
C.11	P-2	23
D.1.1	P-1	24
D.1.2	P-2	24
D.1.3	P-2	25
D.2.1	P-1	26
E.1	P-1	27
E.2	P-1	27
E.3	P-1	27
E.4	P-1	27
E.5	P-1	28

Modern Diesel Technology Job Sheets

for Safety

SAFETY
JOB SHEET #1

Demonstrate Proper Lifting Procedures

Name _____ Station _____ Date _____

NATEF Correlation

This Job Sheet addresses the following NATEF task(s):
None

Performance Objective(s)

Upon completion of this Job Sheet, you will be able to successfully follow the proper procedure when lifting heavy objects.

Tools and Materials

A heavy object with a weight that is within the lifting capability of the technician.
Have various members of the class demonstrate proper weight lifting procedures by lifting an object off the shop floor and placing it on the work bench. Other class members are to observe and record any improper weight lifting procedures.

Protective Clothing/Equipment
None

PROCEDURE

1. If the object is to be carried, be sure your path is free from loose parts or tools.

   Task Completed ____

2. Position your feet close to the object and position your back reasonably straight for proper balance.

   Task Completed ____

3. Your back and elbows should be kept as straight as possible. Continue to bend your knees until your hands reach the best lifting location on the object to be lifted.

   Task Completed ____

4. Be certain the container is in good condition. If a container falls apart during the lifting operation, parts may drop out of the container and result in foot injury or part damage.

   Task Completed ____

5. Maintain a firm grip on the object and do not attempt to change your grip while lifting is in progress.

   Task Completed ____

6. Straighten your legs to lift the object and keep the object close to your body. Use leg muscles rather than back muscles.

   Task Completed ____

7. If you have to change direction of travel, turn your whole body instead of twisting it.

    Task Completed ____

8. Do not bend forward to place an object on a work bench or table. Position the object on the front surface of the work bench and slide it back. Do not pinch your fingers under the object while setting it on the front of the bench.

    Task Completed ____

9. If the object must be placed on the floor or a low surface, bend your legs to lower the object. Do not bend your back forward, because this movement strains back muscles.

    Task Completed ____

10. When a heavy object must be placed on the floor, locate suitable blocks under the object to prevent jamming your fingers under the object.

    Task Completed ____

11. List the improper weight lifting procedures observed during the weight lifting demonstrations.

    1) _____
    2) _____
    3) _____
    4) _____
    5) _____

Problems Encountered

_____
_____
_____

INSTRUCTOR EVALUATION

☐ 4 Mastered Task
☐ 3 Able to Perform Task Independently; Some Additional Training Suggested
☐ 2 Able to Perform Task with Close Supervision; Requires Additional Training
☐ 1 Unable to Perform Task
☐ 0 Not Attempted

Comments

_____
_____

Instructor Name _____ Date _____

Instructor Signature _____

SAFETY
JOB SHEET #2

Operate Vehicles Safely in the Shop

Name _____ Station _____ Date _____

NATEF Correlation
This Job Sheet addresses the following NATEF task(s):
None

Performance Objective(s)
Upon completion of this Job Sheet, you will be able to successfully operate vehicles safely in the shop.

Tools and Materials
A medium or heavy duty truck.
Have a student in the class drive a truck into the shop and park it in a repair bay. Then drive the vehicle out of the shop and park it on the lot. Carefully observe driving procedure and evaluate the procedure using the shop safety driving procedures in steps 1 through 10. List all the improper shop driving procedures that you observed in the spaces at the end of the job sheet.

Protective Clothing/Equipment
None

PROCEDURE

1. Check to be sure no persons or objects are under the truck before you start the engine.

    Task Completed ___

2. Prior to driving a truck, always make sure the brakes are operating and fasten the safety belt.

    Task Completed ___

3. If the truck is equipped with air brakes, be sure the specified air pressure is indicated on the air pressure gauge in the dash before driving the truck.

    Task Completed ___

4. If the truck is parked on a lift, be sure the lift is fully down and the lift arms or components are not contacting the chassis.

    Task Completed ___

5. Check to see if there are any objects directly in front of or behind the truck before driving away.

    Task Completed ___

6. Always drive slowly in the shop and watch carefully for personnel and other moving trucks.

    Task Completed ___

7. Make sure the shop door is high enough so that there is plenty of clearance between the top of the truck and the door.

   Task Completed ___

8. Watch the shop door to be certain that it is not coming down as you attempt to drive under the door.

   Task Completed ___

9. If a road test is necessary, obey all traffic laws, and never drive in a reckless manner.

   Task Completed ___

10. Do not squeal tires when accelerating or turning corners.

    Task Completed ___

11. List any observed improper shop driving procedures.

    1) _____
    2) _____
    3) _____
    4) _____
    5) _____
    6) _____

Problems Encountered

_____
_____
_____

INSTRUCTOR EVALUATION

☐ 4 Mastered Task
☐ 3 Able to Perform Task Independently; Some Additional Training Suggested
☐ 2 Able to Perform Task with Close Supervision; Requires Additional Training
☐ 1 Unable to Perform Task
☐ 0 Not Attempted

Comments

_____
_____

Instructor Name _____ Date _____

Instructor Signature _____

SAFETY
JOB SHEET #3

Shop Housekeeping Inspection

Name _____ Station _____ Date _____

NATEF Correlation
This Job Sheet addresses the following NATEF task(s):
None

Performance Objective(s)
Upon completion of this Job Sheet, you will be able to successfully apply shop housekeeping rules in your shop.

Tools and Materials
When another class of medium/heavy duty truck students is working in the shop, evaluate their shop housekeeping procedures using the shop housekeeping procedures provided in the following list. Record all the improper shop housekeeping procedures that you observed in the spaces at the end of the job sheet.

Protective Clothing/Equipment
None

PROCEDURE

1. Keep aisles and walkways clear of tools, equipment, and other items.

   Task Completed ____

2. Be sure all sewer covers are securely in place.

   Task Completed ____

3. Keep floor surfaces free of oil, grease, water, and loose material.

   Task Completed ____

4. Proper garbage containers must be conveniently located and these containers should be emptied regularly.

   Task Completed ____

5. Access to fire extinguishers must be unobstructed at all times and fire extinguishers should be checked for proper charge at regular intervals.

   Task Completed ____

Warning: When you are finished with a tool, never set it on the customer's truck. After using a tool the best place for it is in your tool box or on the work bench. Many tools have been lost by leaving them on customer's trucks.

6. Tools must be kept clean and in good condition.

                                                                                                   Task Completed ___

7. When not in use, tools must be stored in their proper location.

                                                                        Task Completed ___

8. Oily rags and other combustibles must be placed in proper covered garbage containers.

                                                                        Task Completed ___

9. Rotating components on equipment and machinery must have guards and all shop equipment should have regular service and adjustment schedules.

                                                                        Task Completed ___

10. Benches and seats must be maintained in a clean condition.

                                                                        Task Completed ___

11. Keep parts and materials in their proper locations.

                                                                        Task Completed ___

12. When not in use, creepers must not be left on the shop floor. Creepers should be stored in a specific location.

                                                                        Task Completed ___

13. The shop should be well lighted and all lights should be in working order.

                                                                        Task Completed ___

14. Frayed electric cords on lights or equipment must be replaced.

                                                                        Task Completed ___

15. Walls and windows should be cleaned regularly.

                                                                        Task Completed ___

16. Stairs must be clean, well lighted, and free of loose material.

                                                                        Task Completed ___

17. List any observed improper shop housekeeping.

    1) _____

    2) _____

    3) _____

    4) _____

    5) _____

Problems Encountered

_____

_____

_____

INSTRUCTOR EVALUATION

☐ 4 Mastered Task
☐ 3 Able to Perform Task Independently; Some Additional Training Suggested
☐ 2 Able to Perform Task with Close Supervision; Requires Additional Training
☐ 1 Unable to Perform Task
☐ 0 Not Attempted

Comments

_____

_____

Instructor Name _____ Date _____

Instructor Signature _____

Modern Diesel Technology Job Sheets

for
Electrical/Electronic Systems

ELECTRICAL/ELECTRONIC SYSTEMS
JOB SHEET #1

Wiring Diagrams

Name _____ Station _____ Date _____

NATEF Correlation

This Job Sheet addresses the following NATEF task(s):

A.1 Read, interpret, and diagnose electrical/electronic circuits using wiring diagrams. (P-1)

Performance Objective(s)

Upon completion of this Job Sheet, you will be able to successfully use a wiring diagram as a diagnostic tool.

Tools and Materials

Truck electrical circuit containing examples of opens, shorts and grounded circuits or a circuit board that is similar
Technical service publications containing the necessary diagnostic flow charts and wiring schematics

Protective Clothing/Equipment

Coveralls or shop coat
Safety glasses
Safety shoes

PROCEDURE

1. Locate the wiring schematic that matches the truck or circuit board being used for this task.

   Task Completed ___

2. Select one individual circuit (such as horn circuit, A/C clutch circuit or starter circuit etc.) within the schematic. Using the schematic, trace this circuit.

   Task Completed ___

3. First, find the intended load component (such as a light, motor, solenoid, relay, alarm etc.) Then find the ground for this component.

   Task Completed ___

4. In the space provided below you will now redraw the circuit being traced as a straight line with the component on one end and the battery at the other end. Use the ISO symbols, beginning with the ground connection for the components. Draw the ground symbol and the component symbol.

5. Working backwards from the component toward the battery, redraw the schematic as a straight line in the space below. Using the symbols used in the schematic, indicate the wire color, wire gauge, circuit number, connectors, connector numbers, circuit protection devices, switches, relays and any other components used in this circuit. When the wire color changes (usually at a connector) indicate the new color so that your drawing accurately reflects the circuit. (HINT: Keep your lines short so that your drawing fits in the space provided)

6. Now examine the truck and locate the components used in your drawing. Does your drawing match the circuit on the truck?

Task Completed ___

Problems Encountered

_____
_____
_____

INSTRUCTOR EVALUATION

☐ 4 Mastered Task
☐ 3 Able to Perform Task Independently; Some Additional Training Suggested
☐ 2 Able to Perform Task with Close Supervision; Requires Additional Training
☐ 1 Unable to Perform Task
☐ 0 Not Attempted

Comments

_____
_____
_____

Instructor Name _____   Date _____

Instructor Signature _____

ELECTRICAL/ELECTRONIC SYSTEMS
JOB SHEET #2

Checking for Continuity

Name _____ Station _____ Date _____

NATEF Correlation

This Job Sheet addresses the following NATEF task(s):

A.2 Check continuity in electrical/electronic circuits using appropriate test equipment. (P-1)

A.3 Check applied voltages, circuit voltages, and voltage drops in electrical/electronic circuits using a digital multimeter (DMM). (P-1)

Performance Objective(s)

Upon completion of this Job Sheet, you will be able to successfully use a digital multimeter to check an electrical circuit for continuity.

Tools and Materials
Truck electrical circuit containing examples of opens, shorts and grounded circuits or a circuit board that is similar
Technical service publications containing the necessary diagnostic flow charts and wiring schematics
Digital multimeter
Clamp-on ammeter

Protective Clothing/Equipment
Coveralls or shop coat
Safety glasses
Safety shoes

PROCEDURE

1. Verify the complaint.

   Task Completed ___

2. Using the wiring diagram locate the component and connectors where you can access the circuit to perform voltage tests. Make a list of the test points and indicate what the expected voltage should be at each point in the circuit. The expected voltage will vary depending upon the type of circuit and whether the circuit is power side switched or ground side switched, so be sure to use OEM published specifications.

   _____

   _____

   _____

   _____

3. Switch the circuit on.

   Task Completed ___

4. Connect the test lead connected to the COM port of your multimeter (should be the black one) to a known good ground.

   Task Completed ___

5. Using the schematic, proceed along the circuit testing voltage in reference to ground on each side of each component and connector.

   Task Completed ___

6. When there is a significant or unexpected loss of voltage, perform a voltage drop across the portion of the circuit where the voltage is lost.

   Task Completed ___

7. Compare the voltage drop to OEM published specifications.

   Task Completed ___

8. Replace or repair failed components to restore the circuit function.

   Task Completed ___

9. Verify that the repairs have corrected the complaint.

   Task Completed ___

Problems Encountered

_____
_____
_____

INSTRUCTOR EVALUATION

☐ 4 Mastered Task
☐ 3 Able to Perform Task Independently; Some Additional Training Suggested
☐ 2 Able to Perform Task with Close Supervision; Requires Additional Training
☐ 1 Unable to Perform Task
☐ 0 Not Attempted

Comments

_____
_____
_____

Instructor Name _____ Date _____
Instructor Signature_____

ELECTRICAL/ELECTRONIC SYSTEMS
JOB SHEET #3

Checking Current Flow

Name _____ Station _____ Date _____

NATEF Correlation

This Job Sheet addresses the following NATEF task(s):

A.4 Check current flow in an electrical/electronic circuits and components using a clamp-on ammeter or digital multimeter (DMM). (P-1)

Performance Objective(s)

Upon completion of this Job Sheet, you will be able to successfully use either a digital multimeter (DMM) or clamp-on ammeter to check current flow in electrical/electronic circuits and components.

Tools and Materials
Truck electrical circuit containing examples of opens, shorts and grounded circuits or a circuit board that is similar
Technical service publications containing the necessary diagnostic flow charts and wiring schematics
Digital multimeter
Clamp-on ammeter

Protective Clothing/Equipment
Coveralls or shop coat
Safety glasses
Safety shoes

PROCEDURE

1. Check the amperage measuring limit of your multimeter. This should be printed on the meter and also listed in the meter's manual.

   Task Completed ___

2. Using the technical publications provided, determine whether the expected current flow can safely be tested using the amps or milliamp setting of your multimeter. If in doubt, check the circuit first using the clamp-on ammeter. Larger current flows should be measured using the clamp-on ammeter.

   Task Completed ___

3. Switch the circuit on.

   Task Completed ___

4. Measure very small amounts of current using the digital multimeter connected in series in the circuit. Measure larger amounts of current using the clamp-on ammeter.

   Task Completed ___

5. Is the current measured consistent with published OEM specifications?

   _____

   _____

Problems Encountered

_____
_____
_____

INSTRUCTOR EVALUATION

☐ 4 Mastered Task
☐ 3 Able to Perform Task Independently; Some Additional Training Suggested
☐ 2 Able to Perform Task with Close Supervision; Requires Additional Training
☐ 1 Unable to Perform Task
☐ 0 Not Attempted

Comments

_____
_____
_____

Instructor Name _____ Date _____

Instructor Signature _____

ELECTRICAL/ELECTRONIC SYSTEMS
JOB SHEET #4

Checking Resistance

Name _____ Station _____ Date _____

NATEF Correlation

This Job Sheet addresses the following NATEF task(s):

A.5   Check resistance in electrical/electronic circuits and components using a digital multimeter (DMM). (P-1)

Performance Objective(s)

Upon completion of this Job Sheet, you will be able to successfully measure resistance in a circuit and its components.

Tools and Materials

Truck electrical circuit containing examples of opens, shorts and grounded circuits or a circuit board that is similar
Technical service publications containing the necessary diagnostic flow charts and wiring schematics
Digital multimeter

Protective Clothing/Equipment

Coveralls or shop coat
Safety glasses
Safety shoes

PROCEDURE

1. Using the schematic as a guide, identify several components in the truck electrical circuit or training board that may be removed from the circuit for testing and which have published resistance specifications. Examples would include sensors, relay coils, solenoid coils, diodes, resistors, sending units etc..

    Task Completed ___

2. Remove these components and perform a resistance measurement. When measuring sensors that measure temperature or pressure, refer to the OEM published chart showing the relationship of temperature or pressure to resistance and follow the testing instructions exactly. Does your measurement agree with the specifications for each component?

    _____

    _____

3. Replace the components after testing and verify that they function normally.

    Task Completed ___

4. Using the schematic and the technical manual, identify circuit components which must be tested for resistance while the circuit is operating under load such as relay contacts, solenoid contacts, switch contacts, connectors and wiring etc.. Perform a voltage drop test on each of these components and portions of the circuit.

    Task Completed ___

5. Do the measurements agree with expected voltage drops for these components?

   _____
   _____
   _____

Problems Encountered

_____
_____
_____

INSTRUCTOR EVALUATION

☐ 4 Mastered Task
☐ 3 Able to Perform Task Independently; Some Additional Training Suggested
☐ 2 Able to Perform Task with Close Supervision; Requires Additional Training
☐ 1 Unable to Perform Task
☐ 0 Not Attempted

Comments

_____
_____
_____

Instructor Name _____ Date _____

Instructor Signature_____

ELECTRICAL/ELECTRONIC SYSTEMS
JOB SHEET #5

Opens, Shorts, and Grounds

Name _____ Station _____ Date _____

NATEF Correlation
This Job Sheet addresses the following NATEF task(s):

A.6 Find shorts, grounds, and opens in electrical/electronic circuits. (P-1)

Performance Objective(s)
Upon completion of this Job Sheet, you will be able to successfully diagnose electrical or electronic circuit failures as being either shorted, open, or grounded, and locate the source of the failure.

Tools and Materials
Truck electrical circuit containing examples of opens, shorts and grounded circuits or a circuit board that is similar
Service publication containing the appropriate diagnostic flow charts and wiring schematics.
Digital multimeter
Clamp-on ammeter
Short detector
Calculator

Protective Clothing/Equipment
Coveralls or shop coat
Safety glasses
Safety shoes

PROCEDURE

1. Describe the complaint.

   _____
   _____
   _____

2. Verify the complaint.

   Task Completed ___

3. Perform a visual inspection.

   Task Completed ___

4. Measure the applied voltage and current draw of the circuit.

   Applied Voltage _____

   Current Draw _____

5. Describe the symptoms of a shorted circuit.

_____
_____
_____

6. Describe the symptoms of an open circuit.

_____
_____
_____

7. Describe the symptoms of a grounded circuit

_____
_____
_____

8. Does the complaint resemble the symptoms of an open, shorted, or grounded circuit?

_____

9. Using the technical publication provided, locate the diagnostic flow chart that applies to the fault described. Perform the system checks as directed to locate and repair the problem.

                                                Task Completed ___

10. For problems not covered by a diagnostic flow chart, use the wiring schematic along with a digital multimeter and clamp-on ammeter to locate the circuit fault.

                                                Task Completed ___

Problems Encountered

_____
_____
_____

INSTRUCTOR EVALUATION

☐ 4 Mastered Task
☐ 3 Able to Perform Task Independently; Some Additional Training Suggested
☐ 2 Able to Perform Task with Close Supervision; Requires Additional Training
☐ 1 Unable to Perform Task
☐ 0 Not Attempted

Comments

_____
_____
_____

Instructor Name _____ Date _____

Instructor Signature_____

ELECTRICAL/ELECTRONIC SYSTEMS
JOB SHEET #6

Parasitic Battery Drain Problems

Name _____ Station _____ Date _____

NATEF Correlation

This Job Sheet addresses the following NATEF task(s):

A.7   Diagnose parasitic (key-off) battery drain problems (P-1)

Performance Objective(s)

Upon completion of this Job Sheet, you will be able to successfully measure the amount of parasitic battery drain and locate any problems.

Tools and Materials
Truck electrical system or equivalent circuit board
Wiring diagram
Clamp-on ammeter
Digital multimeter

Protective Clothing/Equipment
Coveralls or shop coat
Safety glasses
Safety shoes

PROCEDURE

1. Switch off all electrical loads, such as lights, and place key switch in the off position. Temporarily prevent the dome light from operating to permit the doors to be opened during the test without causing any current draw.

   Task Completed ___

2. If the truck has multiple batteries, remove any cables connecting the batteries in parallel so that all current must flow through the one cable that is being used for measurements.

   Task Completed ___

3. Ensure that the current draw is less than 5 amps by placing a clamp-on ammeter around the negative battery cable to measure the approximate amount of parasitic load.

   Task Completed ___

4. If parasitic load current is measured as being more than 5A, determine if any device has been left on, such as lights, before proceeding.

   Task Completed ___

5. If no device is found to be switched on, use the clamp-on ammeter and the wiring schematic to trace the circuit. Determine the cause of the excessive parasitic load and correct the problem before proceeding.

    Task Completed ___

6. When parasitic load current is less than 5A, it is now safe to measure the parasitic draw by disconnecting the negative battery cable and connecting a meter set to the milliamp scale in series between the battery terminal and the cable. Observe the meter display for several minutes as electronic devices power down and reset. When the circuit stabilizes record the parasitic draw.

    Parasitic draw is now measured to be _____.

7. Using the wiring diagram count the number of devices such as on-board computers and other devices that need battery power to maintain memory. What is the total amount of power needed for the truck electrical system?

    _____

8. Check in the sleeper compartment for devices such as televisions, microwaves, VCRs, etc. requiring power for their memory. How much additional power is needed?

    _____

9. Is the measured power draw more than the total calculated power draw? If so, locate the source of the excessive draw by pulling fuses, disconnecting circuits, and components until the source of the problem is located.

    Task Completed ___

Problems Encountered

_____

_____

_____

INSTRUCTOR EVALUATION

☐ 4 Mastered Task
☐ 3 Able to Perform Task Independently; Some Additional Training Suggested
☐ 2 Able to Perform Task with Close Supervision; Requires Additional Training
☐ 1 Unable to Perform Task
☐ 0 Not Attempted

Comments

_____

_____

_____

Instructor Name _____ Date _____

Instructor Signature _____

ELECTRICAL/ELECTRONIC SYSTEMS
JOB SHEET #7

Circuit Protection Devices

Name _____ Station _____ Date _____

NATEF Correlation

This Job Sheet addresses the following NATEF task(s):

A.8 Inspect and test fusible links, circuit breakers, relays, solenoids, and fuses; replace as needed. (P-1)

Performance Objective(s)

Upon completion of this Job Sheet, you will be able to successfully inspect and replace as needed, circuit protection devices such as fusible links, circuit breakers, relays, solenoids, and fuses.

Tools and Materials
Digital multimeter
Wiring diagram
Break-out T's for relays and solenoids
Pressure gauges

Protective Clothing/Equipment
Coveralls or shop coat
Safety glasses
Safety shoes

PROCEDURE

1. Remove fuses, circuit breakers and fusible links from the circuit and test their resistance using an ohmmeter. They should have zero or nearly zero resistance. If there is very high or infinite resistance, they are open or tripped. Reset or replace as necessary.

   Task Completed ___

2. To test circuit protection devices while they are installed in the truck, switch on the applicable load and use a digital multimeter to verify the presence of voltage (i.e. + 12V) at both terminals

   Task Completed ___

3. Perform a functionality test of the solenoids and relays by first listening and feeling for an audible click when the relay is actuated. If it does not click when the control circuit is actuated, locate the fault in the control circuit. It may be necessary to use a breakout "T" to perform these measurements

   Task Completed ___

4. If the solenoid/relay clicks when actuated, measure the voltage drop across the load carrying contacts while the circuit is operating. If the voltage drop is excessive, replace the component.

Task Completed ___

5. If the solenoid or relay is being used to control a pressure circuit, perform an additional function check using pressure gauges to verify that the pressure is actually being controlled.

Task Completed ___

Problems Encountered

_____
_____
_____

INSTRUCTOR EVALUATION

☐ 4 Mastered Task
☐ 3 Able to Perform Task Independently; Some Additional Training Suggested
☐ 2 Able to Perform Task with Close Supervision; Requires Additional Training
☐ 1 Unable to Perform Task
☐ 0 Not Attempted

Comments

_____
_____
_____

Instructor Name _____ Date _____

Instructor Signature _____

ELECTRICAL/ELECTRONIC SYSTEMS
JOB SHEET #8

Spike Suppression Devices

Name _____ Station _____ Date _____

NATEF Correlation

This Job Sheet addresses the following NATEF task(s):

A.9   Inspect and test spike suppression diodes/resistors; replace as needed (P-1)

Performance Objective(s)

Upon completion of this Job Sheet, you will be able to successfully locate, identify, test and replace spike suppression diodes or resistors.

Tools and Materials
Truck wiring circuit or equivalent circuit board
Wiring diagram
Digital multimeter
Analog multimeter
Lab scope

Protective Clothing/Equipment
Coveralls or shop coat
Safety glasses
Safety shoes

PROCEDURE

1. Examine the wiring diagram and locate spike suppression diodes/resistors. They may be resistors, diodes, Zener diodes, or capacitors. Be sure to check the schematic of relays and solenoids used in electronic circuits.

    Task Completed ___

2. If they are removable from the circuit, remove them and test individually using the digital multimeter.

    Task Completed ___

3. Often relays and solenoids have a spike suppression device integral with the control circuit. They are connected in parallel with the coil and are not removable. The method of testing these devices varies depending upon which device is being tested. Which device will be used for this test?

    _____

    _____

    _____

4. The test procedure also varies depending upon which type of test equipment is available. It may be an analog ohmmeter, a digital multimeter, or a lab scope. Which type of test equipment will be used for this test?

   _____

5. Describe in detail the steps followed to perform this test.

   _____
   _____
   _____
   _____
   _____
   _____
   _____
   _____
   _____
   _____

6. Perform the test and record the results.

   _____
   _____

7. Replace any faulty components as needed.

   Task Completed ___

Problems Encountered

_____
_____
_____

INSTRUCTOR EVALUATION

☐ 4 Mastered Task
☐ 3 Able to Perform Task Independently; Some Additional Training Suggested
☐ 2 Able to Perform Task with Close Supervision; Requires Additional Training
☐ 1 Unable to Perform Task
☐ 0 Not Attempted

Comments

_____
_____
_____

Instructor Name _____ Date _____
Instructor Signature _____

ELECTRICAL/ELECTRONIC SYSTEMS
JOB SHEET #9

Frequency and Pulse Width

Name _____ Station _____ Date _____

NATEF Correlation

This Job Sheet addresses the following NATEF task(s):

A.10   Check the frequency and pulse width in electrical/electronic circuits using appropriate test equipment. (P-3)

E.1.3   Test headlight and dimmer switch circuit switches, relays, wires terminals, connectors, sockets, and control components; repair or replace as needed (P-1)

Performance Objective(s)

Upon completion of this Job Sheet, you will be able to successfully use several types of test equipment to monitor the frequency and pulse width of various electronic devices.

Tools and Materials
Truck electrical system or equivalent circuit board
Appropriate wiring diagram
Appropriate service literature
Digital multimeter
Lab scope
Electronic Service Tool (EST)

Protective Clothing/Equipment
Coveralls or shop coat
Safety glasses
Safety shoes

PROCEDURE

1. Using the appropriate service literature and wiring diagram identify the devices which may be tested for frequency or duty cycle for this test. Which devices will you be testing?

   _____

2. Describe the procedure recommended for measuring pulse width or frequency for the devices chosen. Indicate which terminals should be tested.

   _____
   _____
   _____

3. Using a digital multimeter configured for measuring frequency, determine the frequency of the signal generated by a variable reluctance type sensor, such as a speed sensor.

   Task Completed ___

4. Using a digital multimeter configured for measuring pulse width, determine the pulse width of a signa,l such as the output of a panel illumination dimmer.

   Task Completed ___

5. Using an oscilloscope, measure the frequency and duty cycle of a signal such as the output of a panel illumination dimmer.

   Task Completed ___

6. Record the measurements.

   _____
   _____
   _____

7. Are the measurements within specifications? _____

Problems Encountered

_____
_____
_____

INSTRUCTOR EVALUATION

☐ 4 Mastered Task
☐ 3 Able to Perform Task Independently; Some Additional Training Suggested
☐ 2 Able to Perform Task with Close Supervision; Requires Additional Training
☐ 1 Unable to Perform Task
☐ 0 Not Attempted

Comments

_____
_____
_____

Instructor Name _____ Date _____

Instructor Signature _____

ELECTRICAL/ELECTRONIC SYSTEMS
JOB SHEET #10

Battery Load Test

Name _____ Station _____ Date _____

NATEF Correlation

This Job Sheet addresses the following NATEF task(s):

B.1 Perform battery load test; determine needed action. (P-1)
B.2 Determine battery state of charge using an open circuit voltage test. (P-1)
B.3 Inspect, clean and service battery, replace as needed (P-1)
B.4 Inspect and clean battery boxes, mounts, and hold downs; repair or replace as needed. (P-1)
B.6 Inspect, test, and clean battery cables and connectors; repair or replace as needed. (P-1)

Performance Objective(s)

Upon completion of this Job Sheet, you will be able to successfully service and replace batteries, cables, and boxes.

Tools and Materials
Truck with a functional electrical system
Digital multimeter
Battery cleaner or baking soda
Battery terminal
Battery charger

Protective Clothing/Equipment
Coveralls or shop coat
Eye protection
Rubber or nitrile gloves

PROCEDURE

1. Inspect the condition of the battery box. Is it securely mounted and does it adequately protect the batteries from the weather and from dirt, moisture and debris?

   Task Completed ___

2. Inspect the battery hold down(s). Are the batteries protected from movement and vibration? Vibration is a significant cause of battery failures.

   Task Completed ___

3. Inspect the condition of the battery cables. Use either a baking soda solution or an equivalent commercial product to clean the cable ends, battery terminals, and the battery cases. Clean the interior of the battery box. Leave the battery cables disconnected from each battery so that they are isolated from each other until after they have been tested and/or recharged.

   Task Completed ___

4. Check the electrolyte level of each cell of conventional and low maintenance batteries. Add distilled water as necessary. Use the internal hydrometer of maintenance free batteries to determine if they have adequate electrolyte.

                                        Task Completed ___

5. Check the state of charge of each battery. Conventional and low maintenance batteries may be checked using a hydrometer or refractometer. *Be sure to temperature correct the specific gravity reading.* Maintenance free batteries can be checked using an open circuit voltage test after any surface charge caused by recent charging has been removed and voltage has stabilized. Recharge and retest any batteries having a low state of charge. WARNING: Do not attempt to recharge a maintenance free battery that has insufficient electrolyte.

                                        Task Completed ___

6. When it is verified that the batteries are sufficiently charged, load test each battery.

                                        Task Completed ___

7. Replace any batteries that have failed. Make sure the batteries are clean and securely mounted. Make sure the battery cables are clean and correctly routed; replace any cables as needed. Reconnect the battery cables.

                                        Task Completed ___

Problems Encountered

_____

_____

_____

INSTRUCTOR EVALUATION

☐ 4 Mastered Task
☐ 3 Able to Perform Task Independently; Some Additional Training Suggested
☐ 2 Able to Perform Task with Close Supervision; Requires Additional Training
☐ 1 Unable to Perform Task
☐ 0 Not Attempted

Comments

_____

_____

_____

Instructor Name _____ Date _____

Instructor Signature _____

ELECTRICAL/ELECTRONIC SYSTEMS
JOB SHEET #11

Battery Charging

Name _____ Station _____ Date _____

NATEF Correlation

This Job Sheet addresses the following NATEF task(s):

B.5 Charge Battery using slow or fast charge method as appropriate (P-1)

Performance Objective(s)

Upon completion of this Job Sheet, you will be able to successfully choose the correct method of charging the battery for the circumstances. You will be able to safely charge the battery correctly.

Tools and Materials
Battery
Battery charger
Battery hydrometer
Refractometer
Battery temperature probe

Protective Clothing/Equipment
Coveralls or shop coat
Safety glasses
Safety shoes
Rubber or nitrile gloves

Warning: Batteries produce an explosive gas while being charged. Do not allow any sparks or flame near batteries. Wear eye and skin protection while working with batteries. Connect the charger to the battery before plugging it into wall current. Never disconnect the charger while it is turned on. Always disconnect the truck ground cable first and reconnect it last. Allow the battery to stabilize before removing the charging cables. Do not attempt to recharge a frozen battery. Make sure to connect the charging cables with the correct polarity. Leave the vent caps in place while the battery is being charged.

PROCEDURE

1. For conventional filler cap batteries and low maintenance batteries, check the electrolyte of the battery and add distilled water if necessary. For sealed maintenance free batteries, examine the built-in hydrometer. Do not recharge the maintenance free battery if the built-in hydrometer indicates that the electrolyte level is low. Sealed batteries with low electrolyte level must be replaced.

   Task Completed ___

2. Disconnect the negative cable and all interconnecting cables to isolate the batteries and to protect any computer or other electronic systems.

   Task Completed ___

3. Measure the state of charge of each battery. For maintenance free batteries, use a digital multimeter to measure the state of charge. For conventional filler cap batteries and low maintenance batteries, measure the state of charge by determining the specific gravity of each cell using a hydrometer. Use a thermometer to measure electrolyte temperature. Use the specific gravity and electrolyte temperature to determine the state of charge for each cell.

   Task Completed ___

4. If there is more than one battery, separate and charge each battery individually.

   Task Completed ___

5. Connect the charger to the battery terminals. Make sure the polarity is correct.

   Task Completed ___

6. Select a rate of charge that does not exceed 30 amps or a voltage higher than 15.0 Volts. A sulfuric acid smell or boiling electrolyte indicates that the battery is being overcharged. Battery temperature should not exceed 125 degrees while charging. Reduce the rate of charge as needed. Where possible use a slow charge to reduce gassing and maximize battery life.

   Task Completed ___

7. For conventional filler cap batteries and low maintenance batteries, use a hydrometer or refractometer to determine when the battery is fully charged. For maintenance free batteries, switch off the battery charger and remove the surface charge using a carbon pile resistor. After removal of the surface charge, use a digital multimeter to measure the open circuit voltage to determine battery state of charge.

   Task Completed ___

8. When the battery is fully charged, turn the charger off, and unplug it. Now disconnect the charger and reconnect the battery cables.

   Task Completed ___

Problems Encountered

_____
_____
_____

INSTRUCTOR EVALUATION

☐ 4 Mastered Task
☐ 3 Able to Perform Task Independently; Some Additional Training Suggested
☐ 2 Able to Perform Task with Close Supervision; Requires Additional Training
☐ 1 Unable to Perform Task
☐ 0 Not Attempted

Comments

_____
_____
_____

Instructor Name _____ Date _____

Instructor Signature_____

ELECTRICAL/ELECTRONIC SYSTEMS
JOB SHEET #12

Jump Starting

Name _____ Station _____ Date _____

NATEF Correlation

This Job Sheet addresses the following NATEF task(s):

B.7 Jump start a vehicle using jumper cables and a booster battery or auxiliary power supply using proper safety procedures. (P-1)

Performance Objective(s)

Upon completion of this Job Sheet, you will be able to successfully jump start a vehicle with a dead battery using either jumper cables or an auxiliary power supply.

Tools and Materials
Vehicle having a dead battery
Jumper cables and booster battery
Auxiliary power supply

Protective Clothing/Equipment
Coveralls or shop coat
Safety glasses
Safety shoes
Rubber or nitrile gloves

Warning: Wear eye protection. Do not attempt to jump start a vehicle with a frozen battery. Whenever possible do not attempt to jump start a vehicle with on-board computer systems. The preferred method is to remove the batteries and recharge them. When it is necessary to jump start vehicles with computer controlled systems make sure that the ignition switch is in the off position in both vehicles before attaching the jumper cables. Make sure the battery has sufficient electrolyte before attempting to jump start. A spark will be produced when the last cable is connected to the vehicle with the dead battery. This connection should be made to the frame or engine block away from the batteries.

PROCEDURE

1. Turn on an electrical load such as the headlights or blower motor of the vehicle with the dead battery before beginning to jump start. This load will absorb any damaging voltage spikes produced while connections are being made.

   Task Completed ___

2. Connect the cables using the following sequence: Connect one end of the positive cable to the positive terminal of the dead battery and then the other end to the positive cable of the booster battery. Connect one end of the negative cable to the negative terminal of the booster battery and then make the final connection by attaching the other end of the negative cable to a good ground, such as the frame or engine block of the vehicle with the dead battery.

   Task Completed ___

3. As soon as the engine starts, disconnect the cables in the reverse order.

                                                                  Task Completed ___

Problems Encountered

_____
_____
_____

INSTRUCTOR EVALUATION

☐ 4 Mastered Task
☐ 3 Able to Perform Task Independently; Some Additional Training Suggested
☐ 2 Able to Perform Task with Close Supervision; Requires Additional Training
☐ 1 Unable to Perform Task
☐ 0 Not Attempted

Comments

_____
_____
_____

Instructor Name _____ Date _____

Instructor Signature_____

ELECTRICAL/ELECTRONIC SYSTEMS
JOB SHEET #13

Battery Capacitance Testing

Name _____ Station _____ Date _____

NATEF Correlation

This Job Sheet addresses the following NATEF task(s):

B.8 Perform battery capacitance test; determine needed action. (P-2)

Performance Objective(s)

Upon completion of this Job Sheet, you will be able to successfully use a capacitance tester to determine a batteries condition and state of charge.

Tools and Materials
Battery
Capacitance tester

Protective Clothing/Equipment
Coveralls or shop coat
Safety glasses
Safety shoes
Rubber or nitrile gloves

Warning: Sparks flame and cigarettes must be kept away from batteries. Always wear protective clothing and eyewear when working with batteries

PROCEDURE

1. Connect the tester to the battery.

    Task Completed ___

2. Enter the information requested by the tester such as whether the testing is being performed in or out of the chassis and the capacity of the battery.

    Task Completed ___

3. Begin the test sequence. It is automated so wait for the test to conclude and the results to be displayed.

    Task Completed ___

Problems Encountered

_____
_____
_____

INSTRUCTOR EVALUATION

☐ 4 Mastered Task
☐ 3 Able to Perform Task Independently; Some Additional Training Suggested
☐ 2 Able to Perform Task with Close Supervision; Requires Additional Training
☐ 1 Unable to Perform Task
☐ 0 Not Attempted

Comments

_____
_____

Instructor Name _____   Date _____

Instructor Signature _____

ELECTRICAL/ELECTRONIC SYSTEMS
JOB SHEET #14

Starter Circuit Testing

Name _____ Station _____ Date _____

NATEF Correlation

This Job Sheet addresses the following NATEF task(s):

C.1 Perform starter circuit cranking voltage and voltage drop tests; determine needed action. (P-1)

Performance Objective(s)

Upon completion of this Job Sheet, you will be able to successfully measure cranking voltage and voltage drops in the starter circuit. Interpretation of these measurements will help determine needed action.

Tools and Materials
Truck with a functional engine and electrical system
Wiring diagram
Digital multimeter
Clamp-on ammeter
Starting/charging system tester with carbon pile

Protective Clothing/Equipment
Coveralls or shop coat
Eye protection
Rubber or nitrile gloves

PROCEDURE

1. Verify the complaint.

   Task Completed ___

2. Measure the state of charge of the batteries. Make sure the batteries are adequately charged and successfully pass a load test before proceeding to test the starting system.

   Task Completed ___

3. Perform a visual inspection of all cables, connections, batteries, and starting system components. Repair any problems you find before proceeding to test the starting system.

   Task Completed ___

4. Disable the fuel system so that the engine will not start during the following test.

   Task Completed ___

5. Connect a voltmeter to measure voltage across the battery terminals during cranking. Crank the engine and record the voltage.

   _____

6. Connect a voltmeter so that it is indicating the voltage between the starter "B" terminal and the starter ground. Crank the engine again and record the voltage during cranking.

   _____

7. Battery voltage and starter voltage should be within 0.5 volts of each other. If the difference is excessive, perform a voltage drop test across the positive battery cable(s), followed by a voltage drop test across the negative battery cable(s) to locate the source of the voltage drop. Describe the results of your test.

_____
_____

8. After the system successfully passes the preceding tests, perform voltage drop tests on the control circuit beginning by measuring the voltage drop between the battery positive post (not the clamp) and the "S" terminal of the cranking motor solenoid. Record the results.

_____
_____

9. Using the wiring diagram as a guide, proceed along the control circuit performing voltage drop tests across portions of the circuit until either the circuit passes or excessive resistance is isolated and repaired.

Task Completed ___

10. When the starting system is performing satisfactorily, restore the fuel system to allow the engine to start; and start the engine to verify the repairs.

Task Completed ___

Problems Encountered

_____
_____
_____

INSTRUCTOR EVALUATION

☐ 4 Mastered Task
☐ 3 Able to Perform Task Independently; Some Additional Training Suggested
☐ 2 Able to Perform Task with Close Supervision; Requires Additional Training
☐ 1 Unable to Perform Task
☐ 0 Not Attempted

Comments

_____
_____
_____

Instructor Name _____   Date _____
Instructor Signature _____

ELECTRICAL/ELECTRONIC SYSTEMS
JOB SHEET #15

Starter Control Circuit

Name _____ Station _____ Date _____

NATEF Correlation

This Job Sheet addresses the following NATEF task(s):

C.2 Inspect and test components (key switch, push button and/or magnetic switch) and wires in the starter control circuit; replace as needed. (P-2)

C.3 Inspect and test, starter relays, and solenoids/switches; replace as needed. (P-2)

Performance Objective(s)

Upon completion of this Job Sheet, you will be able to successfully diagnose and repair problems in the starter control circuit.

Tools and Materials
Truck engine with functional starting circuit
Wiring diagram
Appropriate service publications or textbook
Digital multimeter
Hand tools

Protective Clothing/Equipment
Coveralls or shop coat
Safety glasses
Safety shoes

PROCEDURE

1. Using the wiring diagram, list the components in the starter control circuit.

   _____
   _____
   _____
   _____

2. Using the service publications, calculate the total voltage drop for the starter control circuit by adding amounts permitted for each component.

   _____
   _____
   _____
   _____

3. Using the wiring diagram and the service publication, describe the procedure for measuring the voltage drop of the starter control circuit.

_____
_____
_____
_____

4. Perform voltage drop test of the starter control circuit.

Task Completed ___

5. If your measurement exceeds specifications, test each component and connection until the problem is located.

Task Completed ___

6. Repair or replace the faulty component.

Task Completed ___

Problems Encountered

_____
_____
_____

INSTRUCTOR EVALUATION

☐ 4 Mastered Task
☐ 3 Able to Perform Task Independently; Some Additional Training Suggested
☐ 2 Able to Perform Task with Close Supervision; Requires Additional Training
☐ 1 Unable to Perform Task
☐ 0 Not Attempted

Comments

_____
_____
_____

Instructor Name _____ Date _____

Instructor Signature _____

ELECTRICAL/ELECTRONIC SYSTEMS
JOB SHEET #16

Remove and Replace Starter

Name _____ Station _____ Date _____

NATEF Correlation

This Job Sheet addresses the following NATEF task(s):

C.4 Remove and replace starter; inspect flywheel ring gear or flex plate. (P-2)

Performance Objective(s)

Upon completion of this Job Sheet, you will be able to successfully remove and replace a starter and inspect the condition of the flywheel ring gear or flex plate.

Tools and Materials
Truck engine with functional starting circuit
Wiring diagram
Appropriate service publications
Digital multimeter
Inspection mirror and light
Hand tools

Protective Clothing/Equipment
Coveralls or shop coat
Safety glasses
Safety shoes

PROCEDURE

1. Disconnect the battery cables at the batteries beginning with the negative cable. Wrap the end of the cable with tape to prevent accidental contact with each other.

    Task Completed ___

2. If the truck needs to be lifted to gain working room, lift and safely support the truck.

    Task Completed ___

3. Disconnect and label the wires from the solenoid and the battery cable(s) from the starter.

    Task Completed ___

4. Remove the bolts holding the starter to the engine; make sure the starter is supported before removing the last bolt. Truck starters are heavy and can fall causing injuries. Remove the starter from the chassis.

    Task Completed ___

5. Inspect the flywheel ring gear teeth using the light and mirror. Turn the flywheel one complete revolution to assure that all of the teeth have been examined.

    Task Completed ___

6. Replace the starter and reconnect the wiring and cables at the starter.

                                          Task Completed ___

7. Reconnect the battery cables.

                                          Task Completed ___

8. Perform a functionality test by starting the engine and observing the sound made by the starter motor and its ability to crank the engine.

                                          Task Completed ___

Problems Encountered

_____
_____
_____

INSTRUCTOR EVALUATION

☐ 4 Mastered Task
☐ 3 Able to Perform Task Independently; Some Additional Training Suggested
☐ 2 Able to Perform Task with Close Supervision; Requires Additional Training
☐ 1 Unable to Perform Task
☐ 0 Not Attempted

Comments

_____
_____
_____

Instructor Name _____ Date _____
Instructor Signature_____

ELECTRICAL/ELECTRONIC SYSTEMS
JOB SHEET #17

Charge Indicators

Name _____ Station _____ Date _____

NATEF Correlation

This Job Sheet addresses the following NATEF task(s):

D.1 Diagnose instrument panel mounted volt meters and/or indicator lamps that show no charge, low charge, or overcharge condition; determine needed action. (P-1)

Performance Objective(s)

Upon completion of this Job Sheet, you will be able to successfully diagnose charge indicator devices such as voltmeters or indicator lamps that do not accurately reflect the actual rate of charge.

Tools and Materials
Truck engine with functional charging circuit
Wiring diagram
Appropriate service publications
Digital multimeter
Hand tools

Protective Clothing/Equipment
Coveralls or shop coat
Safety glasses
Safety shoes

PROCEDURE

1. Observe the behavior of the charge indicators as the ignition switch is turned on. If the truck is equipped with an indicator lamp, it should be on; a voltmeter should indicate battery voltage; electronic gauges should perform a self-test with a sweep of the needle.

    Task Completed ___

2. If the indicators do not perform satisfactorily, check the fuses, bulbs and wiring to locate any problems.

    Task Completed ___

3. Connect a digital multimeter to the alternator output terminal.

    Task Completed ___

4. Start the engine and compare the voltage displayed on the multimeter to the dash display, after the output from the alternator has stabilized. If the difference is significant, consult the appropriate service publication for steps to correct the problem.

    Task Completed ___

Problems Encountered

_____
_____
_____

INSTRUCTOR EVALUATION

☐ 4 Mastered Task
☐ 3 Able to Perform Task Independently; Some Additional Training Suggested
☐ 2 Able to Perform Task with Close Supervision; Requires Additional Training
☐ 1 Unable to Perform Task
☐ 0 Not Attempted

Comments

_____
_____
_____

Instructor Name _____ Date _____

Instructor Signature _____

ELECTRICAL/ELECTRONIC SYSTEMS
JOB SHEET #18

Charging System Diagnosis

Name _____ Station _____ Date _____

NATEF Correlation

This Job Sheet addresses the following NATEF task(s):

D.2 Diagnose the cause of no charge, low charge, or overcharge condition; determine needed action (P-1)

Performance Objective(s)

Upon completion of this Job Sheet, you will be able to successfully diagnose the cause of charging system complaints.

Tools and Materials
Truck with a functional electrical system
Wiring diagram this truck
Technical service publications containing the charging system diagnostic flow charts that apply to this system
Digital multimeter
Clamp-on ammeter
Starting/charging system tester with carbon pile

Protective Clothing/Equipment
Coveralls or shop coat
Eye protection
Rubber or nitrile gloves

PROCEDURE

1. Verify the complaint.

   Task Completed ___

2. Perform a preliminary inspection beginning with the drive belt. Check the belt condition, tension, and the function of belt tensioners if equipped.

   Task Completed ___

3. Visually inspect the condition of all charging system related electrical connections including the battery cables.

   Task Completed ___

4. Check battery condition, electrolyte level, and state of charge. If the complaint is overcharging, test for a sulfated battery by performing the three-minute charge test.

   Task Completed ___

5. If the problem has not been located and corrected during the preliminary checks, proceed to check alternator output voltage by connecting a voltmeter between the output terminal and ground. Observe correct polarity of the leads. Start the engine and bring the speed to about 1200 rpm's. Note and record alternator voltage.

_____

6. Now move the voltmeter leads to the battery terminals (the posts, not the cable ends). Observe correct polarity of the leads. With the engine operating at the same speed, note and record battery voltage.

   _____

7. If the battery voltage is more than 0.5 volts lower than the alternator voltage, check and correct the wiring between the alternator output terminal and the battery positive terminal. Also check the ground path between the alternator and the battery negative terminal.

   Task Completed ___

8. If both the alternator and battery voltages are below normal, an undercharge condition exists. Using the diagnostic flow charts provided, perform the pinpoint tests that apply to this system to locate the cause of the problem.

   Task Completed ___

9. If both the alternator and battery voltages are above normal, an overcharging system exists. Using the diagnostic flow charts provided, perform the pinpoint tests that apply to this system to locate the cause of the problem.

   Task Completed ___

Problems Encountered

_____
_____
_____

INSTRUCTOR EVALUATION

☐ 4 Mastered Task
☐ 3 Able to Perform Task Independently; Some Additional Training Suggested
☐ 2 Able to Perform Task with Close Supervision; Requires Additional Training
☐ 1 Unable to Perform Task
☐ 0 Not Attempted

Comments

_____
_____
_____

Instructor Name _____ Date _____

Instructor Signature _____

ELECTRICAL/ELECTRONIC SYSTEMS
JOB SHEET #19

Alternator Belt Drive Systems

Name _____ Station _____ Date _____

NATEF Correlation

This Job Sheet addresses the following NATEF task(s):

D.3 Inspect and replace alternator drive belts, pulleys, fans, tensioners, and mounting brackets; adjust drive belts and check alignment. (P-1)

D.6 Remove and replace alternator. (P-2)

Performance Objective(s)

Upon completion of this Job Sheet, you will be able to successfully maintain the alternator belt drive system.

Tools and Materials
Truck engine with functional charging circuit
Appropriate service publications
Belt tension gauge
Straight edge or steel rule
Hand tools

Protective Clothing/Equipment
Coveralls or shop coat
Safety glasses
Safety shoes

PROCEDURE

1. Inspect the alternator drive belt for cracks, fraying, glazing and edge wear. If it is no longer serviceable it must be replaced.

   Task Completed ___

2. Inspect the pulleys, fans, and idlers for wear.

   Task Completed ___

3. Inspect the tensioners for wear or binding.

   Task Completed ___

4. Inspect the mounting brackets for looseness or cracks.

   Task Completed ___

5. Adjust the belt using the belt tension gauge.

   Task Completed ___

6. Check the alignment of the pulleys using the straight edge.

   Task Completed ___

7. Remove and replace alternator as necessary

Task Completed ___

Problems Encountered

_____
_____
_____

INSTRUCTOR EVALUATION

☐ 4 Mastered Task
☐ 3 Able to Perform Task Independently; Some Additional Training Suggested
☐ 2 Able to Perform Task with Close Supervision; Requires Additional Training
☐ 1 Unable to Perform Task
☐ 0 Not Attempted

Comments

_____
_____
_____

Instructor Name _____   Date _____
Instructor Signature_____

ELECTRICAL/ELECTRONIC SYSTEMS
JOB SHEET #20

Charging Circuit Voltage Drop Testing

Name _____ Station _____ Date _____

NATEF Correlation

This Job Sheet addresses the following NATEF task(s):

D.4 Perform charging system voltage and amperage output tests; determine needed action. (P-1)
D.5 Perform charging circuit voltage drop tests; determine needed action. (P-1)
D.7 Inspect, repair or replace connectors and wires in the charging circuit (P-2)

Performance Objective(s)

Upon completion of this Job Sheet, you will be able to successfully perform voltage drop test of the charging circuit and determine needed action.

Tools and Materials
Truck engine with functional starting circuit
Wiring diagram
Appropriate service publications
Carbon pile or equivalent load device
Digital multimeter
Wire terminals and connecters
Electricians pliers
Hand tools

Protective Clothing/Equipment
Coveralls or shop coat
Safety glasses
Safety shoes

PROCEDURE

1. With the engine not running and the ignition switch in the off position, connect a carbon pile across the alternator output connection and the alternator ground.
   Task Completed ___

2. Place the clamp-on ammeter around the alternator positive output cable so that the direction of conventional current flow is pointing toward the alternator.
   Task Completed ___

3. Connect a digital voltmeter across the alternator positive output connection and alternator ground.
   Task Completed ___

4. Adjust the carbon pile while observing the ammeter to cause the alternator's rated current to flow through the charging circuit.
   Task Completed ___

5. Record the voltage displayed on the voltmeter that is connected across the alternator's positive output terminal and alternator ground.

   Task Completed ___

6. Move the voltmeter so the battery terminal voltage is being measured.

   Task Completed ___

7. Adjust the carbon pile while observing the ammeter to cause the alternator's rated current to flow through the charging circuit.

   Task Completed ___

8. Record the voltage displayed on the voltmeter that is connected across the battery terminals

   Task Completed ___

9. Compare the total charging circuit voltage drop against the manufacturer's specifications. In general, the total charging circuit voltage drop with the alternators rated current flowing should be less than 0.5V (500mV).

   Task Completed ___

10. If the total charging circuit voltage drop is greater than 0.5V, connect a voltmeter between the alternator ground and battery negative terminal and load the carbon pile to cause the alternator's rated current to flow through the charging circuit. Record the voltage dropped on the alternator ground circuit. Repeat this step for the alternator positive circuit. Each reading should be less than the manufacturer's specifications or 0.25V. Voltage drops greater than this indicate excessive resistance in that portion of the charging circuit.

    Task Completed ___

11. Locate and correct the source of the excessive charging circuit resistance such as loose or corroded connections or undersized cables.

    Task Completed ___

12. Inspect and replace any damaged terminals or wires in the charging circuit.

    Task Completed ___

Problems Encountered

_____
_____
_____

INSTRUCTOR EVALUATION

☐ 4 Mastered Task
☐ 3 Able to Perform Task Independently; Some Additional Training Suggested
☐ 2 Able to Perform Task with Close Supervision; Requires Additional Training
☐ 1 Unable to Perform Task
☐ 0 Not Attempted

Comments

_____
_____

Instructor Name _____ Date _____

Instructor Signature_____

ELECTRICAL/ELECTRONIC SYSTEMS
JOB SHEET #21

Leakage of A/C Voltage

Name _____ Station _____ Date _____

NATEF Correlation

This Job Sheet addresses the following NATEF task(s):

D.8 Diagnose A/C voltage leakage (failed rectifier) at alternator output; determine needed action. (P-1)

Performance Objective(s)

Upon completion of this Job Sheet, you will be able to successfully check for the presence of AC voltage leaking from the alternator output.

Tools and Materials
Diesel engine with a complete charging system
Lab scope
Digital multimeter

Protective Clothing/Equipment
Coveralls or shop coat
Safety glasses
Safety shoes

PROCEDURE

1. Set the multimeter to the lowest A/C voltage setting. Attach one lead to the alternator output terminal and the other to the alternator ground.

   Task Completed ___

2. Start the engine and raise to about 1500 rpm's; load the alternator by turning the headlights and also the blower. Observe the level of A/C voltage produced at the alternator output terminal. A reading of over 0.10 VAC may be considered an indication of a failed rectifier.

   Task Completed ___

3. If a lab scope is available, use it to view the waveform while the alternator is loaded. Describe the waveform produced with the waveform expected from a good charging system.

   _____
   _____
   _____
   _____

Problems Encountered

_____
_____
_____

INSTRUCTOR EVALUATION

- ☐ 4 Mastered Task
- ☐ 3 Able to Perform Task Independently; Some Additional Training Suggested
- ☐ 2 Able to Perform Task with Close Supervision; Requires Additional Training
- ☐ 1 Unable to Perform Task
- ☐ 0 Not Attempted

Comments

_____
_____
_____

Instructor Name _____ Date _____

Instructor Signature_____

ELECTRICAL/ELECTRONIC SYSTEMS
JOB SHEET #22

Headlights

Name _____ Station _____ Date _____

NATEF Correlation
This Job Sheet addresses the following NATEF task(s):

Performance Objective(s)
Upon completion of this Job Sheet, you will be able to successfully diagnose and repair failures in the headlight and daytime running light systems; aim the headlights.

Tools and Materials
Truck having a complete headlight system, or an equivalent circuit board
Technical service information
Headlight aiming equipment
Digital multimeter

Protective Clothing/Equipment
Coveralls or shop coat
Safety glasses
Safety shoes

PROCEDURE

1. With the engine off, turn on the headlights and check the low beam, high beam, and high beam indicator in the instrument cluster.

   Task Completed ___

2. Check for intermittent headlights by checking for loose connections and performing a wiggle test of connectors.

   Task Completed ___

3. Check conventionally wired trucks for no lights by checking fuses, power and ground supplies, and the DRL module.

   Task Completed ___

4. Trucks having electronic control of the headlights must be diagnosed using OEM diagnostic procedures. Describe the recommended procedure for the truck chosen for this assignment.

_____

_____

_____

5. Start the engine and allow sufficient time for the charging system voltage to stabilize.

Task Completed ___

6. Check for brighter than normal headlights by checking for low beams that become significantly brighter when engine rpm is increased to about 1500 rpm's. If so, check the charging system for an overcharging condition.

Task Completed ___

7. Check for lights that are too dim by checking voltage available at the light and for a good ground. If voltage is insufficient, locate the source of the lost voltage. If the circuit is satisfactory and the light is still too dim, check for a headlight that has been in service for an extended time.

Task Completed ___

8. Check for intermittent lights by inspecting wiring and connectors. Perform a wiggle test if suspected wiring.

Task Completed ___

9. Using the service information provided, describe how the daytime running lights are controlled. Follow the OEM instructions and tests. Do the DRL's operate normally?

_____

_____

_____

10. Aim the headlights using the equipment provided.

Task Completed ___

Problems Encountered

_____

_____

_____

INSTRUCTOR EVALUATION

☐ 4 Mastered Task
☐ 3 Able to Perform Task Independently; Some Additional Training Suggested
☐ 2 Able to Perform Task with Close Supervision; Requires Additional Training
☐ 1 Unable to Perform Task
☐ 0 Not Attempted

Comments

_____
_____
_____

Instructor Name _____ Date _____
Instructor Signature_____

ELECTRICAL/ELECTRONIC SYSTEMS
JOB SHEET #23

Lighting Components

Name _____ Station _____ Date _____

NATEF Correlation

This Job Sheet addresses the following NATEF task(s):

E.1.3 Test Headlight and dimmer circuit switches relays, wires, terminals, connectors, sockets and control components; repair or replace as needed. (P1)

E.1.4 Inspect and test switches, bulbs/LEDs, sockets, connectors, terminals, relays, and wires of parking, clearance, and taillights circuits; repair or replace as needed. (P-1)

E.1.5 Inspect and test instrument panel light circuit switches relays, bulbs, sockets, connectors, terminals, wires and printed circuits/control modules; repair or replace as needed. (P-2)

E.1.6 Inspect and test interior cab light circuit switches, bulbs, sockets, connectors, terminals and wires; repair or replace as needed. (P-2)

E.2.1 Inspect, test, and adjust stoplight circuit switches, bulbs/LEDs, sockets, connectors, terminals, and wires; repair or replace as needed. (P-1)

E.2.2 Inspect and test turn signal and hazard circuit flasher(s), switches relays, bulbs/LEDs, , sockets, connectors, terminals, wires; repair or replace as needed. (P-1)

E.2.3 Inspect, test, and adjust backup lights and warning device circuit switches, bulbs/LEDs, sockets, horns, buzzers, connectors, terminals, and wires; repair or replace as needed. (P-2)

F.5 Inspect and test warning devices (lights and audible) circuit sending units. bulbs/LEDs, sockets, connectors, wires and printed circuits/control modules; repair or replace as needed. (P-2)

Performance Objective(s)

Upon completion of this Job Sheet, you will be able to successfully inspect, repair and maintain the components found in lighting and warning circuits.

Tools and Materials

Truck with a functional electrical system or an equivalent circuit board(s)
Wiring diagram
Service publications
Digital multimeter
Wire terminals, sockets, connectors, switches, bulbs/LEDs, and wiring

Protective Clothing/Equipment

Coverall or shop coat
Safety glasses
Safety shoes

PROCEDURE

1. Verify the complaint.

   Task Completed ___

2. Using the technical service information provided, describe how the affected circuit is controlled and locate the power and ground circuits. Please use space on top of next page.

_____

_____

_____

_____

3. If the circuit is multiplexed or controlled by an electronic control module, follow the OEM instructions and steps in their diagnostic flow charts to test the system and locate the source of the failure.

    Task Completed ___

4. Perform preliminary checks beginning with a visual inspection and a functionality check of suspected components.

    Task Completed ___

5. Inspect terminals, sockets, and connectors for corrosion, loose connections, backed out terminals, and signs of overheating. Repair or replaced as needed.

    Task Completed ___

6. Perform a voltage drop test of switches, relays, connectors, terminals and wires to locate excessive resistance. Repair or replace as needed.

    Task Completed ___

7. Diagnosis of printed circuits, electronic control modules and multiplexed systems can only be performed by following the instructions and performing the tests as instructed by the OEM's. If the complaint being used for this task requires this, describe what you did to find the problem.

_____

_____

_____

_____

Problems Encountered

_____

_____

_____

INSTRUCTOR EVALUATION

☐ 4 Mastered Task
☐ 3 Able to Perform Task Independently; Some Additional Training Suggested
☐ 2 Able to Perform Task with Close Supervision; Requires Additional Training
☐ 1 Unable to Perform Task
☐ 0 Not Attempted

Comments

_____

_____

_____

Instructor Name _____ Date _____

Instructor Signature _____

ELECTRICAL/ELECTRONIC SYSTEMS
JOB SHEET #24

7-way Trailer Cord

Name _____ Station _____ Date _____

NATEF Correlation

This Job Sheet addresses the following NATEF task(s):

E.1.7 Inspect and test tractor-to-trailer multi-wire connector(s); repair or replace as needed. (P-1)

Performance Objective(s)

Upon completion of this Job Sheet, you will be able to successfully test and maintain the tractor-to-trailer cord.

Tools and Materials
Truck and trailer with a functional lighting system or an equivalent circuit board
Appropriate wiring diagram
7-Way cord tester
Trailer cord brush
Hand tools
Digital multimeter

Protective Clothing/Equipment
Coveralls or shop coat
Safety glasses
Safety shoes

PROCEDURE

1. Use a wire brush to clean the terminal sockets in the end of the trailer cord.

    Task Completed ___

2. Connect the 7-way trailer cord tester to the end of the trailer cord.

    Task Completed ___

3. Check the operation of the left and right turn signals.

    Task Completed ___

4. Check the operation of the hazard lights.

    Task Completed ___

5. Check the operation of the stop lights.

    Task Completed ___

6. Check the operation of the tail lights.

    Task Completed ___

7. Check the operation of the marker lights.

    Task Completed ___

Problems Encountered

_____
_____
_____

INSTRUCTOR EVALUATION

- ☐ 4 Mastered Task
- ☐ 3 Able to Perform Task Independently; Some Additional Training Suggested
- ☐ 2 Able to Perform Task with Close Supervision; Requires Additional Training
- ☐ 1 Unable to Perform Task
- ☐ 0 Not Attempted

Comments

_____
_____
_____

Instructor Name _____   Date _____

Instructor Signature _____

ELECTRICAL/ELECTRONIC SYSTEMS
JOB SHEET #25

Using On-board Computer Self Diagnostics

Name _____ Station _____ Date _____

NATEF Correlation

This Job Sheet addresses the following NATEF task(s):

F.1 Interface with vehicle's on-board computer; perform diagnostic procedure using recommended electronic diagnostic equipment and tools (including PC based software and/or data scan tools); determine needed action. (P-1)

Performance Objective(s)

Upon completion of this Job Sheet, you will be able to successfully access the truck's self diagnostics using an Electronic Service Tool (EST).

Tools and Materials
Truck with a functional electrical system that is equipped with on-board diagnostics and an ATA data connector
Technical manuals or on-line data containing the service information for this truck
Wiring diagram
EST with the correct software cartridge or PC with diagnostic software
Data cables
Digital multimeter

Protective Clothing/Equipment
Coveralls or shop coat
Eye protection
Rubber or nitrile gloves

Warning: There is a wide variety of Electronic Service Tools and PC based Software versions. The following steps are intended only to provide a generic description of how the TYPICAL equipment is used. Before you attempt to complete this job-sheet, read the operating instructions for the equipment you will be using for this exercise. Be sure you understand and follow the instructions and procedures completely before proceeding. If you are not sure; ask your instructor.

PROCEDURE

1. Make sure that the cables and software are the correct ones for this truck.

   Task Completed ___

2. Locate the data connector. It should be located inside the cab on the left side of the steering column. Connect the cable and the EST to the ATA connector.

   Task Completed ___

3. Turn the ignition switch on. Wait for the EST to establish communication with the vehicle diagnostics. When communication is verified, follow the on-screen prompts. You may be prompted to select the MID associated with the system you wish to access.

   Task Completed ___

4. Start the engine and select the data list from the menu. Observe that data is being displayed and is reflecting engine operating conditions. This confirms that there is communication. On a separate sheet of paper make a list of the data displayed. Record the values (temperatures, pressures etc...) displayed.

   Task Completed ___

5. Select active diagnostic codes. On a separate sheet of paper record any codes present.

   Task Completed ___

6. Select Logged Fault Codes. On a separate sheet of paper, record any codes present. If your system supports PID-FMI codes record them. If your system supports it, record the number of occurrences, the first occurrences, the last occurrence and the time on the diagnostic clock for each code.

   Task Completed ___

7. Select Logged Event Codes. Make a list of the event codes displayed along with the number of occurrences, and the first and last occurrence.

   Task Completed ___

8. Select the Trip Recorder and from the Trip Recorder menu select Driver Trip Data. Make a copy of the data displayed

   Task Completed ___

9. Shut off the engine, disconnect the data cable and compare your notes to normal operating conditions. What are your conclusions?

   _____
   _____
   _____

Problems Encountered

_____
_____
_____

INSTRUCTOR EVALUATION

- ☐ 4 Mastered Task
- ☐ 3 Able to Perform Task Independently; Some Additional Training Suggested
- ☐ 2 Able to Perform Task with Close Supervision; Requires Additional Training
- ☐ 1 Unable to Perform Task
- ☐ 0 Not Attempted

Comments

_____
_____
_____

Instructor Name _____ Date _____

Instructor Signature _____

Modern Diesel Technology Job Sheets

for
Heating, Ventilation & Air Conditioning

HEATING, VENTILATION, & AIR CONDITIONING
JOB SHEET #1

Unusual Operating Noises

Name _____ Station _____ Date _____

NATEF Correlation

This Job Sheet addresses the following NATEF task(s):

A.1 Verify the need for service or repair of HVAC systems based on unusual operating noises; determine needed action. (P-1)

Performance Objective(s)

Upon completion of this Job Sheet, you will be able to successfully recognize unusual noises associated with HVAC system operation.

Tools and Materials
Truck with a functional HVAC system or equivalent mock-up

Protective Clothing/Equipment
Coveralls or shop coat
Safety glasses
Safety shoes
Rubber or nitrile gloves

PROCEDURE

1. Turn the key on but do not start the engine yet, turn the blower to its lowest speed and listen, turn the blower to each of the other speeds while listening for unusual noises. Describe any problems observed.

   _____
   _____

2. Start the engine and note the sound. Now turn the controls to the maximum A/C setting. Did you hear the A/C compressor clutch engage? Describe the sound produced by the compressor while it is operating. Does the compressor cycle on and off? If so, how often does it cycle? Describe any problems found.

   _____
   _____
   _____

3. Based on these steps, what are your conclusions?

_____
_____
_____

Problems Encountered

_____
_____
_____

INSTRUCTOR EVALUATION

☐ 4 Mastered Task
☐ 3 Able to Perform Task Independently; Some Additional Training Suggested
☐ 2 Able to Perform Task with Close Supervision; Requires Additional Training
☐ 1 Unable to Perform Task
☐ 0 Not Attempted

Comments

_____
_____
_____

Instructor Name _____ Date _____
Instructor Signature _____

HEATING, VENTILATION, & AIR CONDITIONING
JOB SHEET #2

Unusual Operating Conditions

Name _____ Station _____ Date _____

NATEF Correlation
This Job Sheet addresses the following NATEF task(s):

Performance Objective(s)
Upon completion of this Job Sheet, you will be able to successfully observe the operation of an HVAC system and determine if there are any unusual sights, odors, or temperatures.

Tools and Materials
Truck with a functional HVAC system or equivalent mock-up

Protective Clothing/Equipment
Coveralls or shop coat
Safety glasses
Safety shoes
Rubber or nitrile gloves

PROCEDURE

1. Is this HVAC system computer controlled and networked to the data bus? If so, the first step in troubleshooting should be to connect an EST to the chassis data bus.

   Task Completed ___

2. Select MID 190 for the HVAC system.

   Task Completed ___

3. Scroll through the self diagnostic display and list any fault codes present.

   Task Completed ___

4. List the Failure Mode Indicator (FMI) for each code.

   Task Completed ___

5. What diagnostic tips and tests are recommended?

   _____
   _____
   _____

6. Start the engine and bring it to normal operating temperature. Move the temperature control to the highest temperature setting. Describe the amount of heat produced at the outlets. Which outlets produce heat?

    _____
    _____
    _____
    _____

7. Change the controls to defrost. Describe the amount of heat coming from the defroster outlets. Is there an odor with the defroster operating?

    _____
    _____
    _____

8. Were there any unusual smells while the heater was running?

    _____

9. Change the control to maximum A/C setting; allow several minutes for the system to stabilize. Describe the temperature of the air coming from the outlets.

    _____

10. Were there any unusual odors while the A/C was running?

    _____

11. Did the windshield fog up during your tests?

    _____

12. Based on your observations, what are your conclusions?

    _____
    _____
    _____
    _____

Problems Encountered

_____
_____
_____

INSTRUCTOR EVALUATION

☐ 4 Mastered Task
☐ 3 Able to Perform Task Independently; Some Additional Training Suggested
☐ 2 Able to Perform Task with Close Supervision; Requires Additional Training
☐ 1 Unable to Perform Task
☐ 0 Not Attempted

Comments

_____
_____
_____

Instructor Name _____ Date _____
Instructor Signature_____

HEATING, VENTILATION, & AIR CONDITIONING
JOB SHEET #3

Performance Testing of Different Types of HVAC Systems

Name _____ Station _____ Date _____

NATEF Correlation

This Job Sheet addresses the following NATEF task(s):

A.3   Identify system type and components (cycling clutch orifice tube - CCOT, expansion valve) and conduct performance test(s) on HVAC systems; determine needed action. (P-1)

Performance Objective(s)

Upon completion of this Job Sheet, you will be able to successfully identify the type of HVAC system and conduct a performance test.

Tools and Materials
Truck with a functional HVAC system or equivalent mock-up
Appropriate service publications
Temperature probe or thermometer

Protective Clothing/Equipment
Coveralls or shop coat
Safety glasses
Safety shoes
Rubber or nitrile gloves

PROCEDURE

1. Refer to the service publication that applies to this exercise. Make a list of the components used in this system.

   _____
   _____
   _____
   _____
   _____
   _____

2. Open the hood and examine the A/C system. Verify that the components on the truck match your list; modify your list as necessary. Does this system have an expansion valve or an orifice tube; does it have an accumulator or a receiver/drier? What type of system is this?

   _____
   _____
   _____

3. Locate the under hood label and examine the type of service ports used with this system. What type of refrigerant is used in this system? What is the quantity of the refrigerant charge?

   _____
   _____

4. Place the thermometer in the center dash outlet. Set the A/C control to the coldest temperature setting. Turn the fan to its highest speed.

   Task Completed ___

5. Start the engine; observe the gauges until they reach normal operating conditions. Raise the engine speed to 1200 to 1400 RPM's. Operate the system for at least 5 minutes to stabilize it, record the temperature on the thermometer. What is it? An efficient system should produce air that is at least 30 degrees colder than ambient air.

   _____

6. Based upon your observations, is the system operating satisfactorily?

   _____

Problems Encountered

_____
_____
_____

INSTRUCTOR EVALUATION

☐ 4 Mastered Task
☐ 3 Able to Perform Task Independently; Some Additional Training Suggested
☐ 2 Able to Perform Task with Close Supervision; Requires Additional Training
☐ 1 Unable to Perform Task
☐ 0 Not Attempted

Comments

_____
_____
_____

Instructor Name _____   Date _____

Instructor Signature _____

HEATING, VENTILATION, & AIR CONDITIONING
JOB SHEET #4

Temperature Control Problems

Name _____ Station _____ Date _____

NATEF Correlation
This Job Sheet addresses the following NATEF task(s):

Performance Objective(s)
Upon completion of this Job Sheet, you will be able to successfully service and maintain the temperature control system.

Tools and Materials
Truck with a functional HVAC system or equivalent mock-up
Service publications containing diagnostic flow charts and wiring schematics
Temperature probe or thermometer
Pressure/vacuum gauge
Hand tools

Protective Clothing/Equipment
Coveralls or shop coat
Safety glasses
Safety shoes
Rubber or nitrile gloves

PROCEDURE

1. Is this HVAC system computer controlled and networked to the data bus? If so, the first step in troubleshooting should be to connect an EST to the chassis data bus.

   Task Completed ___

2. Select MID 190 for the air HVAC system.

   Task Completed ___

3. Scroll through the self diagnostic display and list any fault codes present.

   Task Completed ___

4. List the Failure Mode Indicator (FMI) for each code.

   Task Completed ___

5. What diagnostic tips and tests are recommended?

   _____
   _____
   _____

6. Note what happens when you start the engine and then turn the A/C controls to their maximum setting. Did the compressor clutch engage? You should hear the clutch engage, and notice a change in engine RPM and sound. If you are unable to tell by this method, observe the compressor clutch while an assistant operates the controls. If the clutch fails to engage, check the wiring schematic for the system to identify the power and ground sources and any compressor control devices used. Refer to the service publications and follow the manufacturer's diagnostic flow chart to pinpoint the source of the failure. Repair any problems found before proceeding.

_____
_____

7. When you have verified the compressor clutch functions normally; perform a quick check to see if the A/C system is also operating normally. Compare the compressor suction and discharge temperatures. The suction side should be cool to the touch and the discharge should be hot. Describe your results.

_____
_____

8. With the air conditioning operating normally, check the systems ability to direct and blend air to the desired temperature and outlet. The temperature of HVAC system output air is generally controlled by one of two ways. 1) In blend air systems, air cooled by the evaporator core is mixed with air warmed by passing through the heater core. Ultimately, the output air temperature control occurs by regulating the amount of air allowed to flow through each core. 2) In other A/C systems, temperature control is achieved by cycling the compressor clutch. Operate the controls and note whether you are able to direct air to the desired outlet and change the temperature as desired. If the system fails, proceed to the service publication and follow the appropriate diagnostic routine. Describe what you did.

_____
_____
_____

Problems Encountered

_____
_____
_____

INSTRUCTOR EVALUATION

☐ 4 Mastered Task
☐ 3 Able to Perform Task Independently; Some Additional Training Suggested
☐ 2 Able to Perform Task with Close Supervision; Requires Additional Training
☐ 1 Unable to Perform Task
☐ 0 Not Attempted

Comments

_____
_____

Instructor Name _____   Date _____

Instructor Signature_____

HEATING, VENTILATION, & AIR CONDITIONING
JOB SHEET #5

A/C System Pressure and Temperature Problems

Name _____ Station _____ Date _____

NATEF Correlation

This Job Sheet addresses the following NATEF task(s):

B.1.3 Diagnose A/C system problems indicated by pressure gauge and temperature readings; determine needed action. (P-1)

B.3.10 Identify and inspect A/C system service ports (gauge connections); determine needed action. (P-1)

Performance Objective(s)

Upon completion of this Job Sheet, you will be able to successfully connect a manifold gauge set to an A/C system and take pressure and temperature measurements. Diagnose system failures based upon your measurements.

Tools and Materials
Truck with a functional HVAC system, or equivalent mock-up
Appropriate service publications
A/C charging station or manifold gauge set
Temperature probe or thermometer
Hand tools

Protective Clothing/Equipment
Coveralls or shop coat
Safety glasses
Safety shoes
Rubber or nitrile gloves

PROCEDURE

1. Is this HVAC system computer controlled and networked to the data bus? If so, the first step in troubleshooting should be to connect an EST to the chassis data bus.

    Task Completed ___

2. Select MID 190 for the HVAC system.

    Task Completed ___

3. Scroll through the self diagnostic display and list any fault codes present.

    Task Completed ___

4. List the Failure Mode Indicator (FMI) for each code.

    Task Completed ___

5. What diagnostic tips and tests are recommended?

    _____

    _____

6. Identify the type of refrigerant used in the system by referring to the under hood label and checking the type of service port fittings that are unique to each type of refrigerant. What type of refrigerant is used in this system?

   _____
   _____

7. Select the correct manifold gauge set or service hoses.

   Task Completed ___

8. Ensure the low side hand valve is fully closed - turned fully clockwise. Ensure that the high side hand valve is also fully closed - turned fully clockwise.

   Task Completed ___

9. Remove the protective cap from both the low and high side service ports. Make sure that both hose end valves are closed. Connect the low side service hose coupler to the low side port and the high side coupler to the high side port.

   Task Completed ___

10. Open the hose end valves.

    Task Completed ___

11. Record the static pressure of the system before starting the engine and engaging the compressor clutch.

    _____

12. Start the engine and conduct a performance test of the system. Record your temperature and pressure readings.

    _____
    _____
    _____

13. What should the temperature and pressures be?

    _____
    _____
    _____

14. What are your conclusions?

    _____
    _____
    _____

Problems Encountered

_____
_____
_____

INSTRUCTOR EVALUATION

☐ 4 Mastered Task
☐ 3 Able to Perform Task Independently; Some Additional Training Suggested
☐ 2 Able to Perform Task with Close Supervision; Requires Additional Training
☐ 1 Unable to Perform Task
☐ 0 Not Attempted

Comments

_____
_____
_____

Instructor Name _____   Date _____
Instructor Signature_____

HEATING, VENTILATION, & AIR CONDITIONING
JOB SHEET #6

Unusual Operating Noises

Name _____ Station _____ Date _____

NATEF Correlation

This Job Sheet addresses the following NATEF task(s):

B.1.4 Diagnose A/C system problems indicated by visual, audible, smell and touch procedures; determine needed action. (P-1)

Performance Objective(s)

Upon completion of this Job Sheet, you will be able to successfully observe the operation of an A/C system and diagnose any unusual operating conditions.

Tools and Materials
Truck with a functional HVAC system or equivalent mock-up
Appropriate service publications
Temperature probe or thermometer
Hand tools

Protective Clothing/Equipment
Coveralls or shop coat
Safety glasses
Safety shoes
Rubber or nitrile gloves

PROCEDURE

1. Describe the type of system being examined.

   _____
   _____

2. Begin a performance test of the system (during the performance test you will be examining the system while it is operating). Describe the steps you will take.

   _____
   _____
   _____

3. While the system is operating, listen to the sound of the compressor clutch, compressor, blower motor(s), and the engine. Describe your observations.

   _____
   _____
   _____

4. Examine the hoses and fittings for signs of leakage; examine the compressor shaft seal for signs of leakage. Describe your observations.

   _____
   _____

5. Examine the condenser for signs of debris or blockage to air flow; check the condition of the mounts and hose connections. Check the evaporator drain. Describe your findings.

   _____
   _____

6. Using a temperature probe or your hands compare the temperature of the low pressure side of the system at several points along the hoses and lines with the temperature of the high side of the system at several locations, such as the discharge side of the compressor and both sides of the condenser. What is the temperature of the air discharged from the outlets? Describe your findings.

   _____
   _____
   _____

7. Are there any unusual odors?

   _____

8. What are your conclusions?

   _____
   _____

Problems Encountered

_____
_____
_____

INSTRUCTOR EVALUATION

☐ 4 Mastered Task
☐ 3 Able to Perform Task Independently; Some Additional Training Suggested
☐ 2 Able to Perform Task with Close Supervision; Requires Additional Training
☐ 1 Unable to Perform Task
☐ 0 Not Attempted

Comments

_____
_____
_____

Instructor Name _____ Date _____

Instructor Signature _____

HEATING, VENTILATION, & AIR CONDITIONING
JOB SHEET #7

A/C System Leak Testing

Name _____ Station _____ Date _____

NATEF Correlation

This Job Sheet addresses the following NATEF task(s):

B.1.5 Perform A/C system leak test; determine needed action. (P-1)

Performance Objective(s)

Upon completion of this Job Sheet, you will be able to successfully use either electronic leak detection equipment or fluorescent dye to locate the source of refrigerant leaks.

Tools and Materials

Truck with a functional HVAC system or equivalent mock-up
Appropriate service publications
Electronic leak detector
Fluorescent dye leak detector
Hand tools

Protective Clothing/Equipment

Coveralls or shop coat
Safety glasses
Safety shoes
Rubber or nitrile gloves

PROCEDURE

1. Inspect the system looking for oily residue along the hoses, at connections, and around sealing points. A common source of leaks is the service ports and their caps. Over-tightened hose connections can also be distorted and leak. The compressor seal is another common source of leaks. Any places showing evidence of leakage will be examined more closely using leak detection equipment. Describe any problems found.

   _____
   _____
   _____
   _____

2. If an electronic leak detector is provided, begin by calibrating the tester. Then use it to check suspected areas of leakage. Because gaseous refrigerants are heavier than air, move the probe slowly underneath suspected leaks. Continue until the leak source is identified.

   Task Completed ___

3. If fluorescent dye leak detection equipment is available, install the dye following the instructions that pertain to your equipment. Perform a quick check by running the system for 15 minutes and checking for leaks. If none are evident, recheck the system after it has operated for at least 24 hours. Continue to examine the system until the leak is located.

   Task Completed ___

Problems Encountered

_____
_____
_____

INSTRUCTOR EVALUATION

☐ 4 Mastered Task
☐ 3 Able to Perform Task Independently; Some Additional Training Suggested
☐ 2 Able to Perform Task with Close Supervision; Requires Additional Training
☐ 1 Unable to Perform Task
☐ 0 Not Attempted

Comments

_____
_____
_____

Instructor Name _____ Date _____

Instructor Signature _____

HEATING, VENTILATION, & AIR CONDITIONING
JOB SHEET #8

Evacuate A/C System

Name _____ Station _____ Date _____

NATEF Correlation

This Job Sheet addresses the following NATEF task(s):

B.1.6 Evacuate A/C system using appropriate equipment. (P-1)

Performance Objective(s)

Upon completion of this Job Sheet, you will be able to successfully recover the refrigerant from the system and evacuate the system to remove moisture.

Tools and Materials

Truck with a functional HVAC system or equivalent mock-up
Appropriate service publications
A/C recovery and charging station (or similar equipment) with instructions included
Hand tools

Protective Clothing/Equipment

Coveralls or shop coat
Safety glasses
Safety shoes
Rubber or nitrile gloves

PROCEDURE

1. Locate the service ports. Describe their location. What type of refrigerant is used in the system? If the type of refrigerant is in question, use a refrigerant identifier to verify the presence of pure refrigerant.

   _____
   _____

2. If the system uses R-134a, make sure the hose end valves are closed before proceeding. Connect the hoses to the service ports. Now open the hose end valves and follow instructions to recover the refrigerant.

   Task Completed ___

3. After the refrigerant is successfully recovered, turn the controls to begin evacuating using the recovery station vacuum pump. Observe and record the action of the low and high side gauges as the system begins evacuating. Describe your observations.

   _____
   _____
   _____

4. After 5 minutes of evacuating, the low side gauge should indicate a vacuum of about 20 in Hg (67.6 kPa) and the high side gauge should read less that zero (unless your gauge restricts its movement with a stop in). If the high side does not drop to a reading of less than zero there, may be a blockage in the system. If there is a blockage, locate and repair the problem before proceeding. Describe your findings.

_____
_____
_____

5. Continue evacuating. After 15 minutes, the low side gauge should have stabilized and should indicate about 24 to 29.9 in Hg (depending upon your altitude). If it reads less than this value, close the low side valve and observe to gauge. If the low side needle rises slowly, there is a leak in the system. Repair the leak before proceeding. If there are no leaks, continue to evacuate for at least 30 minutes. Longer evacuation time may be needed at higher altitudes and at lower ambient temperatures. Consult your service publications for the optimum procedure. Describe your actions.

_____
_____
_____

6. When there are no leaks and evacuation is complete, the system is ready to be recharged.

Task Completed ___

Problems Encountered

_____
_____
_____

INSTRUCTOR EVALUATION

☐ 4 Mastered Task
☐ 3 Able to Perform Task Independently; Some Additional Training Suggested
☐ 2 Able to Perform Task with Close Supervision; Requires Additional Training
☐ 1 Unable to Perform Task
☐ 0 Not Attempted

Comments

_____
_____
_____

Instructor Name _____ Date _____

Instructor Signature _____

HEATING, VENTILATION, & AIR CONDITIONING
JOB SHEET #9

Contaminated A/C System Cleaning

Name _____ Station _____ Date _____

NATEF Correlation

This Job Sheet addresses the following NATEF task(s):

B.1.7 Internally clean contaminated A/C system components and hoses. (P-2)

Performance Objective(s)

Upon completion of this Job Sheet, you will be able to successfully clean the A/C system internally.

Tools and Materials
Truck with a functional HVAC system or equivalent mock-up
Appropriate service publications
A/C flushing equipment
A/C evacuating and recharging equipment
Hand tools

Protective Clothing/Equipment
Coveralls or shop coat
Safety glasses
Safety shoes
Rubber or nitrile gloves

PROCEDURE

1. The system must be cleaned internally by purging or flushing whenever there has been a compressor failure, a ruptured desiccant bag, water contamination or the presence of unknown refrigerants, oils, or dyes. Describe the reason the system is being flushed.

   _____
   _____
   _____

2. A variety of methods are used to clean a contaminated system. The most recommended methods of cleaning the system are 1) Solvent flushing, older R-12 systems only 2) Refrigerant flushing 3) Installation of in-line filters and 4) Replacing contaminated components. Because some systems should not be flushed, be sure to check the manufacturer's recommendations. Which method(s) you will be performing?

   _____
   _____
   _____

3. The steps you must follow for this assignment will vary for each of the methods recommended. Describe the method to be used to complete this assignment and list the steps you will be performing.

_____
_____
_____
_____
_____
_____
_____
_____

4. Has the system been successfully cleaned?

Task Completed ___

Problems Encountered

_____
_____
_____

INSTRUCTOR EVALUATION

☐ 4 Mastered Task
☐ 3 Able to Perform Task Independently; Some Additional Training Suggested
☐ 2 Able to Perform Task with Close Supervision; Requires Additional Training
☐ 1 Unable to Perform Task
☐ 0 Not Attempted

Comments

_____
_____
_____

Instructor Name _____ Date _____

Instructor Signature_____

HEATING, VENTILATION, & AIR CONDITIONING
JOB SHEET #10

Charging A/C System

Name _____ Station _____ Date _____

NATEF Correlation

This Job Sheet addresses the following NATEF task(s):

B.1.8 Charge A/C system with refrigerant. (P-1)

Performance Objective(s)

Upon completion of this Job Sheet, you will be able to successfully charge the system with refrigerant.

Tools and Materials
Truck with a functional HVAC system or equivalent mock-up
Appropriate service publications
Refrigerant charging station with instructions included
Refrigerant
Temperature probe or thermometer
Hand tools

Protective Clothing/Equipment
Coveralls or shop coat
Safety glasses
Safety shoes
Rubber or nitrile gloves

PROCEDURE

1. The system should have been previously been evacuated and leak tested. Has this been accomplished?

   Task Completed ___

2. Adding the correct level of refrigerant to the system is extremely critical. The durability of the system and its efficiency depend upon having the correct charge. Overcharged systems will develop high head pressures and temperatures; undercharged systems lead to superheating the refrigerant in the low side causing high compressor temperatures and inadequate lubrication. What is the correct size of the refrigerant charge for this system?

   _____

3. Connect the refrigerant container to the service hose.

   Task Completed ___

4. Open the service hose shut-off valve.

   Task Completed ___

5. Check the system for blockage by briefly opening the high side hand valve, allowing some pressure into the system. Observe the low side gauge. If it does not promptly move from a vacuum into a pressure range, a blockage is indicated. Repair any blockage found before proceeding.

   Task Completed ___

6. Open the low side and high side shut-off valves.

   Task Completed ___

7. If your equipment measures and injects the refrigerant with the engine off, select the correct amount of refrigerant for the system. Use the charging equipment to install the refrigerant. If your equipment requires it, measure the correct refrigerant weight and install in the system now.

   Task Completed ___

8. Close the service hose valve.

   Task Completed ___

9. Charging is complete when the correct weight of refrigerant has entered the system. Make sure that there is not liquid refrigerant in the compressor by turning it over by hand several revolutions.

   Task Completed ___

10. Now start the engine and begin a system performance test to verify that the system is operating efficiently. Is the system operating normally?

    _____

11. Close the valves and remove the equipment.

    Task Completed ___

12. Open the low and high side manifold hose shut-off valves.

    Task Completed ___

13. Replace all protective caps and covers.

    Task Completed ___

Problems Encountered

_____
_____
_____

INSTRUCTOR EVALUATION

☐ 4 Mastered Task
☐ 3 Able to Perform Task Independently; Some Additional Training Suggested
☐ 2 Able to Perform Task with Close Supervision; Requires Additional Training
☐ 1 Unable to Perform Task
☐ 0 Not Attempted

Comments

_____
_____
_____

Instructor Name _____ Date _____

Instructor Signature_____

HEATING, VENTILATION, & AIR CONDITIONING
JOB SHEET #11

Identify Lubricant Type Needed for System Application

Name _____ Station _____ Date _____

NATEF Correlation

This Job Sheet addresses the following NATEF task(s):

B.1.9 Identify lubricant type needed for system application. (P-1)

Performance Objective(s)

Upon completion of this Job Sheet, you will be able to successfully use a refrigerant identifier.

Tools and Materials
Truck with a functional HVAC system or equivalent mock-up
Appropriate service publications
Graduated container

Protective Clothing/Equipment
Coveralls or shop coat
Safety glasses
Safety shoes
Rubber or nitrile gloves

PROCEDURE

1. Identify the type of refrigerant used in the system by examining the type of service ports and checking the under-hood label. If there is any doubt, use a refrigerant identifier to verify what is in the system. Describe.

   _____

2. This is the description of how to use a "TYPICAL" refrigerant identifier. Your equipment may vary slightly. Before connecting the hose to the vehicle or tank, power up the refrigerant identifier.

   Task Completed ___

3. The refrigerant identifier will now perform a calibration process that may take several minutes. Some models will prompt to enter the altitude. When the tester prompts that it is ready THEN connect the hose to the vehicle or tank of refrigerant to be identified.

   Task Completed ___

4. After the hose is connected, press the key when prompted by the tester to begin the analysis. The tester will now begin testing a sample of the refrigerant. When it is finished, "R-12", or" R-134a" or "UNKNOWN" will be displayed. Some models will also indicate amount of air in the refrigerant.

   Task Completed ___

5. If "UNKNOWN" is displayed, do not recover this refrigerant into the tank of pure refrigerant. It must be recovered into a separate tank labeled "UNKNOWN" for disposal or to be reclaimed by further processing. DO NOT recycle or reuse this refrigerant. *Caution: propane is commonly used as refrigerant in Mexico. Be especially careful when working with Mexican reefer trailers.*

   Task Completed ___

6. What refrigerant does this system use?

   _____

   _____

7. The type of refrigeration oil recommended for each system varies. Using the service publications provided, identify the correct lubricant.

   Task Completed ___

8. What is the correct lubricant for this system?

   _____

Problems Encountered

_____

_____

_____

INSTRUCTOR EVALUATION

☐ 4 Mastered Task
☐ 3 Able to Perform Task Independently; Some Additional Training Suggested
☐ 2 Able to Perform Task with Close Supervision; Requires Additional Training
☐ 1 Unable to Perform Task
☐ 0 Not Attempted

Comments

_____

_____

_____

Instructor Name _____ Date _____

Instructor Signature _____

HEATING, VENTILATION, & AIR CONDITIONING
JOB SHEET #12

A/C System Protection Devices

Name _____ Station _____ Date _____

NATEF Correlation

This Job Sheet addresses the following NATEF task(s):

B.2.1 Diagnose A/C system problems that cause protection devices (pressure, thermal, and electronic) to interrupt system operation; determine needed action. (P-1)

Performance Objective(s)

Upon completion of this Job Sheet, you will be able to successfully diagnose a condition that causes the system protection devices to interrupt operation.

Tools and Materials
Truck with a functional HVAC system or equivalent mock-up
Appropriate service publications
Temperature probe or thermometer
Manifold gauge set or A/C charging station
Digital multimeter (DMM)
Hand tools

Protective Clothing/Equipment
Coveralls or shop coat
Safety glasses
Safety shoes
Rubber or nitrile gloves

PROCEDURE:

1. Is this HVAC system computer controlled and networked to the data bus? If so, the first step in troubleshooting should be to connect an EST to the chassis data bus.

   Task Completed ___

2. Select MID 190 for the air HVAC system.

   Task Completed ___

3. Scroll through the self diagnostic display and list any fault codes present.

   Task Completed ___

4. List the Failure Mode Indicator (FMI) for each code.

   Task Completed ___

5. What diagnostic tips and tests are recommended?

   _____
   _____
   _____

6. A variety of A/C system protection devices are used to interrupt compressor operation to prevent compressor damage. If the system loses refrigerant to dangerously low levels, a low pressure switch opens and prevents compressor engagement. Other devices are used to interrupt compressor operation if system pressures rise too high. Some systems have a high pressure relief valve that discharges excessively high pressures. Refer to the service publication provided and list the circuit protection devices used in this system.

   _____
   _____

7. Describe the conditions that would cause each of the devices on your list to interrupt compressor operation.

   _____
   _____
   _____

8. List any control modules responsible for managing your system.

   _____

9. Connect a gauge set for your refrigerant service station to the system.

   Task Completed ___

10. Observe static pressure with the system not operating. Is the pressure high enough (above 28 PSI) to close the low pressure cut-out switch? If the pressure is too low, locate and repair the leak and then evacuate and recharge the system. Retest when the system is properly charged. If static pressure is high enough but the compressor does not engage, use the multimeter and the wiring diagram to locate a possible electrical failure. What caused the compressor clutch to fail to engage?

    _____
    _____
    _____

11. High pressure cut-out devices are used to protect systems from excessive pressures. Check your system for excessive high pressures.

    Task Completed ___

12. If the system was interrupted by excessive high pressures, check for conditions causing the high pressures such as: restricted condenser airflow, refrigerant overcharge, air in the system, etc.. Correct as needed.

    Task Completed ___

13. Some systems have a compressor temperature switch that monitors compressor temperatures and opens to protect the compressor. Did the compressor clutch disengage due to excessive compressor temperatures?

    _____

Problems Encountered

_____
_____
_____

INSTRUCTOR EVALUATION

☐ 4 Mastered Task
☐ 3 Able to Perform Task Independently; Some Additional Training Suggested
☐ 2 Able to Perform Task with Close Supervision; Requires Additional Training
☐ 1 Unable to Perform Task
☐ 0 Not Attempted

Comments

_____
_____
_____

Instructor Name _____ Date _____

Instructor Signature _____

HEATING, VENTILATION, & AIR CONDITIONING
JOB SHEET #13

A/C Compressor Belt Drive System

Name _____ Station _____ Date _____

NATEF Correlation

This Job Sheet addresses the following NATEF task(s):

B.2.3 Inspect and replace A/C compressor drive belts, pulleys, and tensioners; adjust belt tension and check alignment. (P-1)

Performance Objective(s)

Upon completion of this Job Sheet, you will be able to successfully service and maintain the compressor belt drive system.

Tools and Materials
Truck with a functional HVAC system or equivalent mock-up
Appropriate service publications
Hand tools

Protective Clothing/Equipment
Coveralls or shop coat
Safety glasses
Safety shoes
Rubber or nitrile gloves

PROCEDURE

1. It is important to ensure the compressor belts pulleys and tensioners are in good condition. Inspect the belt for being loose, cracked, glazed, frayed, or having edge wear. Describe the condition of the belt.

   _____
   _____

2. Check the pulleys and idlers for wear and/or rough bearings. Describe their condition.

   _____
   _____

3. If there is a tensioner, check it for binding and ensure it is providing the correct belt tension. If the belt tension is manually adjusted, check and adjust its tension.

   Task Completed ___

4. A belt with frayed edges indicates misalignment of the pulleys. Use a straight edge, steel rule or laser device to check alignment of the pulleys. Correct as needed

   Task Completed ___

Problems Encountered

_____

_____

_____

INSTRUCTOR EVALUATION

☐ 4 Mastered Task
☐ 3 Able to Perform Task Independently; Some Additional Training Suggested
☐ 2 Able to Perform Task with Close Supervision; Requires Additional Training
☐ 1 Unable to Perform Task
☐ 0 Not Attempted

Comments

_____

_____

_____

Instructor Name _____   Date _____

Instructor Signature _____

HEATING, VENTILATION, & AIR CONDITIONING
JOB SHEET #14

A/C Compressor

Name _____ Station _____ Date _____

NATEF Correlation

This Job Sheet addresses the following NATEF task(s):

B.2.6 Inspect, test, and replace A/C compressor. (P-2)
B.2.7 Inspect, repair, or replace A/C compressor mountings and hardware. (P-2)

Performance Objective(s)

Upon completion of this Job Sheet, you will be able to successfully service and replace the compressor and its hardware.

Tools and Materials

Truck with a functional HVAC system or equivalent mock-up
Appropriate service publications
Hand tools

Protective Clothing/Equipment

Coveralls or shop coat
Safety glasses
Safety shoes
Rubber or nitrile gloves

PROCEDURE

1. Disconnect the negative battery cable.

   Task Completed ___

2. Label and disconnect electrical connections to the compressor.

   Task Completed ___

3. Identify and recover the refrigerant.

   Task Completed ___

4. Disconnect the refrigerant lines or manifold to the compressor; cap the openings.

   Task Completed ___

5. Remove the compressor; drain and measure the refrigerant oil in the compressor.

   Task Completed ___

6. Drain the refrigerant oil from the replacement compressor; turn the compressor several turns by hand to insure that all of the oil is removed.

   Task Completed ___

7. Measure and add the specified amount of refrigerant oil to the replacement compressor; turn the compressor several times by hand to insure that there is not oil above the pistons.

    Task Completed ___

8. Install the new compressor; check the alignment of the belts.

    Task Completed ___

9. Uncap and install the hoses; reconnect the electrical connections.

    Task Completed ___

10. Evacuate and recharge the system.

    Task Completed ___

11. Conduct a performance test of the system.

    Task Completed ___

Problems Encountered

_____
_____
_____

INSTRUCTOR EVALUATION

☐ 4 Mastered Task
☐ 3 Able to Perform Task Independently; Some Additional Training Suggested
☐ 2 Able to Perform Task with Close Supervision; Requires Additional Training
☐ 1 Unable to Perform Task
☐ 0 Not Attempted

Comments

_____
_____
_____

Instructor Name _____ Date _____
Instructor Signature _____

HEATING, VENTILATION, & AIR CONDITIONING
JOB SHEET #15

Lubricant Level

Name _____ Station _____ Date _____

NATEF Correlation

This Job Sheet addresses the following NATEF task(s):

B.3.1 Correct system lubricant level when replacing the evaporator, condenser, receiver/drier or accumulator/drier, and hoses. (P-1)

Performance Objective(s)

Upon completion of this Job Sheet, you will be able to successfully maintain the correct system lubricant level when system components are replaced.

Tools and Materials

Truck with a functional HVAC system or equivalent mock-up
Appropriate service publications
A/C lubricant
Hand tools

Protective Clothing/Equipment

Coveralls or shop coat
Safety glasses
Safety shoes
Rubber or nitrile gloves

PROCEDURE

1. Drain and measure the refrigerant oil from the component being replaced; compare the amount of oil recovered with the amount specified for that component.

   Task Completed ___

2. Measure and add the specified amount of new refrigerant oil to the new component.

   Task Completed ___

Problems Encountered

_____
_____
_____

INSTRUCTOR EVALUATION

☐ 4 Mastered Task
☐ 3 Able to Perform Task Independently; Some Additional Training Suggested
☐ 2 Able to Perform Task with Close Supervision; Requires Additional Training
☐ 1 Unable to Perform Task
☐ 0 Not Attempted

Comments

_____

_____

Instructor Name _____ Date _____

Instructor Signature _____

HEATING, VENTILATION, & AIR CONDITIONING
JOB SHEET #16

Lines, Hoses, Fittings and Seals

Name _____ Station _____ Date _____

NATEF Correlation

This Job Sheet addresses the following NATEF task(s):

B.3.2 Inspect A/C system hoses, lines, filters, fittings, and seals; determine needed action. (P-1)

Performance Objective(s)

Upon completion of this Job Sheet, you will be able to successfully service A/C system hoses, lines, filters, fittings, and seals.

Tools and Materials

Truck with a functional HVAC system or equivalent mock-up
Appropriate service publications
Hand tools

Protective Clothing/Equipment

Coveralls or shop coat
Safety glasses
Safety shoes
Rubber or nitrile gloves

PROCEDURE

1. You should check A/C system hoses for damage and leaks during the course of any A/C system maintenance or inspection. Hoses should be replaced if they are cracked, kinked, abraded, or if the fittings show any signs of abuse. Describe the condition of the hoses.

   _____
   _____
   _____

2. Recover the refrigerant. *Caution: Do not discharge refrigerant to atmosphere. It is both illegal and also environmentally irresponsible.*

   Task Completed ___

3. Disassemble leaking fittings and replace any damaged or leaking O-rings. New O-rings should be lubricated with the appropriate refrigeration oil during installation.

   Task Completed ___

3. A/C system filters and screens are used to prevent particulate (from corrosion, compressor failure, or desiccant breakdown) from circulating through the A/C system and must be replaced if they are clogged, restricted, or damaged. Some systems have a filter in the line between the condenser and the evaporator and some of these filters contain an orifice tube. You must install this type of filter in the proper direction.

   Task Completed ___

Problems Encountered

_____
_____
_____

INSTRUCTOR EVALUATION

☐ 4 Mastered Task
☐ 3 Able to Perform Task Independently; Some Additional Training Suggested
☐ 2 Able to Perform Task with Close Supervision; Requires Additional Training
☐ 1 Unable to Perform Task
☐ 0 Not Attempted

Comments

_____
_____
_____

Instructor Name _____   Date _____
Instructor Signature_____

HEATING, VENTILATION, & AIR CONDITIONING JOB SHEET #17

A/C Condenser Maintenance

Name _____ Station _____ Date _____

NATEF Correlation

This Job Sheet addresses the following NATEF task(s):

B.3.3 Inspect A/C condenser for proper air flow. (P-1)
B.3.4 Inspect and test A/C system condenser and mountings; determine needed action. (P-2)

Performance Objective(s)

Upon completion of this Job Sheet, you will be able to successfully check the condenser for adequate air flow.

Tools and Materials

Truck with a functional HVAC system or equivalent mock-up
Appropriate service publications
Temperature probe or thermometer
Fin comb
Hand tools

Protective Clothing/Equipment

Coveralls or shop coat
Safety glasses
Safety shoes
Rubber or nitrile gloves

PROCEDURE

1. Inspect the condenser for restrictions to airflow. Carefully clean away accumulations of insects, dirt, grime, and debris. Do not use a high pressure washer that can damage the fins.

   Task Completed ___

2. Use a fin comb to straighten any damaged fins.

   Task Completed ___

3. Inspect condenser mounts and repair as needed.

   Task Completed ___

Problems Encountered

_____
_____
_____

INSTRUCTOR EVALUATION

☐ 4 Mastered Task
☐ 3 Able to Perform Task Independently; Some Additional Training Suggested
☐ 2 Able to Perform Task with Close Supervision; Requires Additional Training
☐ 1 Unable to Perform Task
☐ 0 Not Attempted

Comments

_____
_____
_____

Instructor Name _____ Date _____

Instructor Signature _____

HEATING, VENTILATION, & AIR CONDITIONING JOB SHEET #18

Receiver/Drier and Accumulator/Drier Service

Name _____ Station _____ Date _____

NATEF Correlation

This Job Sheet addresses the following NATEF task(s):

B.3.5 Inspect and replace receiver/drier or accumulator/drier. (P-1)

Performance Objective(s)

Upon completion of this Job Sheet, you will be able to successfully replace the systems receiver/drier or accumulator/drier.

Tools and Materials
Truck with a functional HVAC system or equivalent mock-up
Appropriate service publications
Receiver/drier or accumulator/drier, as required
Refrigerant
Refrigerant oil
A/C charging station
Hand tools

Protective Clothing/Equipment
Coveralls or shop coat
Safety glasses
Safety shoes
Rubber or nitrile gloves

PROCEDURE

1. Recover the refrigerant in the system.

   Task Completed ___

2. Disconnect the negative battery cable.

   Task Completed ___

3. Disconnect the electrical connections to any switches that are attached.

   Task Completed ___

4. Disconnect and cap the hoses at both the inlet and outlet of the receiver/drier.

   Task Completed ___

5. Remove the mounting bolts and brackets necessary for removal of the unit.

   Task Completed ___

6. Remove the receiver/drier or accumulator/drier from the truck.

   Task Completed ___

7. Replace and lubricate the O-rings.

Task Completed ___

8. Position the new unit and install the bolts and brackets.

Task Completed ___

9. Uncap and reconnect the hoses.

Task Completed ___

10. Transfer any switches and reconnect.

Task Completed ___

11. Evacuate and recharge the system.

Task Completed ___

12. Performance test the system.

Task Completed ___

Problems Encountered

_____
_____
_____

INSTRUCTOR EVALUATION

☐ 4 Mastered Task
☐ 3 Able to Perform Task Independently; Some Additional Training Suggested
☐ 2 Able to Perform Task with Close Supervision; Requires Additional Training
☐ 1 Unable to Perform Task
☐ 0 Not Attempted

Comments

_____
_____
_____

Instructor Name _____ Date _____
Instructor Signature _____

HEATING, VENTILATION, & AIR CONDITIONING
JOB SHEET #19

Orifice Tube Replacement

Name _____ Station _____ Date _____

NATEF Correlation

This Job Sheet addresses the following NATEF task(s):

B.3.7 Inspect and replace orifice tube. (P-1)

Performance Objective(s)

Upon completion of this Job Sheet, you will be able to successfully inspect and replace the orifice tube.

Tools and Materials
Truck with a functional HVAC system or equivalent mock-up
Appropriate service publications
A/C charging station
Orifice tube
Hand tools

Protective Clothing/Equipment
Coveralls or shop coat
Safety glasses
Safety shoes
Rubber or nitrile gloves

PROCEDURE

1. Recover the refrigerant.

   Task Completed ___

2. Disconnect the negative battery cable.

   Task Completed ___

3. Disconnect and cap the inlet line at the evaporator.

   Task Completed ___

4. Pour a small amount of refrigerant oil into the orifice tube.

   Task Completed ___

5. Insert the orifice tube removal tool and turn the handle until the tool engages the tabs of the orifice tube. Remove the orifice tube.

   Task Completed ___

6. Lubricate the new orifice tube with clean refrigerant oil. Place the tube in the evaporator inlet and push until it is seated.

        Task Completed ___

7. Install a new O-ring on the refrigerant line and reconnect.

        Task Completed ___

8. Evacuate and recharge the system.

        Task Completed ___

9. Performance test the system.

        Task Completed ___

Problems Encountered

_____
_____
_____

INSTRUCTOR EVALUATION

☐ 4 Mastered Task
☐ 3 Able to Perform Task Independently; Some Additional Training Suggested
☐ 2 Able to Perform Task with Close Supervision; Requires Additional Training
☐ 1 Unable to Perform Task
☐ 0 Not Attempted

Comments

_____
_____
_____

Instructor Name _____ Date _____
Instructor Signature_____

HEATING, VENTILATION, & AIR CONDITIONING
JOB SHEET #20

Evaporator Housing

Name _____ Station _____ Date _____

NATEF Correlation

This Job Sheet addresses the following NATEF task(s):

B.3.9 Inspect, clean, and repair evaporator housing and water drain; inspect and service/replace evaporator air filter. (P-1)

Performance Objective(s)

Upon completion of this Job Sheet, you will be able to successfully service the evaporator housing.

Tools and Materials
Truck with a functional HVAC system or equivalent mock-up
Appropriate service publications
A/C charging station
Evaporator housing cleaning compound
Evaporator housing filter
Hand tools

Protective Clothing/Equipment
Coveralls or shop coat
Safety glasses
Safety shoes
Rubber or nitrile gloves

PROCEDURE

1. Locate the evaporator drain tube. Locate the evaporator drain tube for the sleeper compartment if so equipped. Examine the drain(s) for obstructions. Describe your findings.

   _____

   _____

2. Disconnect the drain(s) and clean away any debris or contamination. What did you find?

   _____

   _____

3. Remove the blower(s) and examine the accessible area of the evaporator case for debris or contamination. Clean thoroughly. Were you able to clean away all of the contamination?

   _____

4. If the evaporator case needs additional cleaning to remove bacteria and mold, use a commercial cleaning agent to clean the evaporator case. Is the evaporator case sufficiently cleaned?

   _____

5. If the evaporator case needs additional repairs it will be necessary to remove the evaporator. Begin by recovering the refrigerant.

   Task Completed ___

6. Remove the evaporator.

   Task Completed ___

7. Perform the repairs or cleaning necessary to restore the evaporator case to service.

   Task Completed ___

8. Replace the evaporator.

   Task Completed ___

9. Evacuate and recharge the system.

   Task Completed ___

10. Performance test the system.

    Task Completed ___

Problems Encountered

_____
_____
_____

INSTRUCTOR EVALUATION

☐ 4 Mastered Task
☐ 3 Able to Perform Task Independently; Some Additional Training Suggested
☐ 2 Able to Perform Task with Close Supervision; Requires Additional Training
☐ 1 Unable to Perform Task
☐ 0 Not Attempted

Comments

_____
_____
_____

Instructor Name _____   Date _____

Instructor Signature _____

HEATING, VENTILATION, & AIR CONDITIONING
JOB SHEET #21

Window Fogging Problems

Name _____ Station _____ Date _____

NATEF Correlation

This Job Sheet addresses the following NATEF task(s):

C.2   Diagnose window fogging problems; determine needed action (P-2)

Performance Objective(s)

Upon completion of this Job Sheet, you will be able to successfully diagnose window fogging problems and determine any needed action.

Tools and Materials

Truck with a functional HVAC system or equivalent mock-up
Service publications containing diagnostic flow charts
Temperature probe or thermometer
Hand tools

Protective Clothing/Equipment

Coveralls or shop coat
Safety glasses
Safety shoes
Rubber or nitrile gloves

PROCEDURE

1. Start the engine; bring to operating temperature.

    Task Completed ___

2. Select defrost. Select the highest fan speed. Measure the temperature of the air produced at the defroster outlets.

    _____

3. Describe any odors such as coolant or mildew produced by the defroster.

    _____

4. Use the diagnostic charts provided in the service publications provided to determine any needed action. What repairs are necessary?

    _____

    _____

Problems Encountered

_____
_____
_____

INSTRUCTOR EVALUATION

☐ 4 Mastered Task
☐ 3 Able to Perform Task Independently; Some Additional Training Suggested
☐ 2 Able to Perform Task with Close Supervision; Requires Additional Training
☐ 1 Unable to Perform Task
☐ 0 Not Attempted

Comments

_____
_____
_____

Instructor Name _____   Date _____

Instructor Signature_____

HEATING, VENTILATION, & AIR CONDITIONING JOB SHEET #22

Engine Cooling System

Name _____ Station _____ Date _____

NATEF Correlation

This Job Sheet addresses the following NATEF task(s):

C.3 Perform engine cooling system tests for leaks, protection level, contamination, coolant level, coolant type, temperature, and conditioner concentration; determine needed action. (P-1)

C.4 Inspect engine cooling and heating system hoses, lines, and clamps; determine needed action. (P-1)

C.5 Inspect and test radiator, pressure cap and coolant recovery system (surge tank); determine needed action. (P-1)

Performance Objective(s)

Upon completion of this Job Sheet, you will be able to successfully maintain the engine cooling system.

Tools and Materials
Truck with a functional HVAC system or equivalent mock-up
Appropriate service publications
Refractometer
Hydrometer
Coolant test strips
TDS probe
Cooling system pressure tester
Temperature probe or thermometer
Hand tools

Protective Clothing/Equipment
Coveralls or shop coat
Safety glasses
Safety shoes
Rubber or nitrile gloves

PROCEDURE

1. With the engine stopped and coolant at ambient temperature, check the engine coolant level using the sight glass if provided or the cold level mark on the surge tank.

   Task Completed ___

2. Identify the type of coolant used.

   Task Completed ___

3. Using test strips check the conditioner concentration.

   Task Completed ___

4. Using the hydrometer or refractometer provided, measure the antifreeze protection level.

   Task Completed ___

5. Take a sample of the coolant and check for contamination such as dissolved solids (using a TDS probe), oil, or fuel. Describe what is found.

   _____
   _____

6. Using a cooling system pressure tester, test the cooling system. Check all hoses, lines and clamps for leaks. If a hose is leaking or bulged on the outside, check the inside of the hose for cracks. What actions are needed?

   _____

7. Pressure test the radiator cap. Is the recovery system operating normally?

   _____

8. Start the engine and bring to operating temperature. Check the temperature of the coolant.

   Task Completed ___

9. Based on your measurements and observations, what action is needed?

   _____
   _____
   _____

Problems Encountered

_____
_____
_____

INSTRUCTOR EVALUATION

☐ 4 Mastered Task
☐ 3 Able to Perform Task Independently; Some Additional Training Suggested
☐ 2 Able to Perform Task with Close Supervision; Requires Additional Training
☐ 1 Unable to Perform Task
☐ 0 Not Attempted

Comments

_____
_____
_____

Instructor Name _____ Date _____

Instructor Signature _____

HEATING, VENTILATION, & AIR CONDITIONING
JOB SHEET #23

Recover, Flush, and Refill Cooling System

Name _____ Station _____ Date _____

NATEF Correlation

This Job Sheet addresses the following NATEF task(s):

C.8 Recover, flush and refill with recommended coolant/additive package; bleed cooling system. (P-1)

C.11 Inspect and flush heater core; determine needed action. (P-2)

Performance Objective(s)

Upon completion of this Job Sheet, you will be able to successfully recover, flush and refill the cooling system with the recommended coolant/additive package.

Tools and Materials
Truck with a functional HVAC system or equivalent mock-up
Appropriate service publications
Coolant recovery/flushing station
Coolant test strips
Refractometer
Hydrometer
Hand tools

Protective Clothing/Equipment
Coveralls or shop coat
Safety glasses
Safety shoes
Rubber or latex gloves

PROCEDURE

1. Following the instructions accompanying the coolant recovery/flushing station attach the unit to the truck cooling system.

   Task Completed ___

2. Recover the coolant.

   Task Completed ___

3. Flush the radiator.

   Task Completed ___

4. Flush the heater core(s).

   Task Completed ___

5. Flush the block.

   Task Completed ___

6. Replace the coolant filter, if so equipped.

          Task Completed ___

7. Refill the system with the recommended coolant mixture and additive. Bleed air from the system to insure a complete fill.

          Task Completed ___

8. Remove the recovery/flushing station.

          Task Completed ___

9. Start the engine and bring to operating temperature.

          Task Completed ___

10. Check for leaks and for proper operation.

          Task Completed ___

Problems Encountered

_____
_____
_____

INSTRUCTOR EVALUATION

☐ 4 Mastered Task
☐ 3 Able to Perform Task Independently; Some Additional Training Suggested
☐ 2 Able to Perform Task with Close Supervision; Requires Additional Training
☐ 1 Unable to Perform Task
☐ 0 Not Attempted

Comments

_____
_____
_____

Instructor Name _____ Date _____

Instructor Signature_____

HEATING, VENTILATION, & AIR CONDITIONING
JOB SHEET #24

HVAC Electrical Control Systems

Name _____ Station _____ Date _____

NATEF Correlation

This Job Sheet addresses the following NATEF task(s):
D.1.1 Diagnose the cause of failures in HVAC electrical control systems; determine needed action. (P-1)
D.1.2 Inspect and test A/C heater blower motors, resistors, switches, relays, modules, wiring and protection devices; determine needed action. (P-2)

Performance Objective(s)

Upon completion of this Job Sheet, you will be able to successfully diagnose the cause of failures in the HVAC electrical and electronic control systems.

Tools and Materials
Truck with a functional HVAC system or equivalent mock-up
Appropriate service publications
Digital multimeter
Electronic Service Tool (EST)
Hand tools

Protective Clothing/Equipment
Coveralls or shop coat
Safety glasses
Safety shoes
Rubber or nitrile gloves

PROCEDURE

1. Is this HVAC system computer controlled and networked to the data bus? If so, the first step in troubleshooting should be to connect an EST to the chassis data bus.

   Task Completed ___

2. Select MID 190 for the HVAC system.

   Task Completed ___

3. Scroll through the self diagnostic display and list any fault codes present.

   Task Completed ___

4. List the Failure Mode Indicator (FMI) for each code.

   Task Completed ___

5. What diagnostic tips and tests are recommended?

   _____

   _____

6. Turn on the ignition switch. Move the control through each of the available fan speeds. Which of the available speeds are present?

   _____

7. Listen to the sound of the fan motor. If the fan is noisy, perform any repairs needed.

   Task Completed ___

8. If one or more of the available fan speeds is absent, follow the diagnostic flow chart in the service manual to locate and correct the problem.

   Task Completed ___

9. If the fan does not operate, verify that adequate voltage is present at the blower motor and that it has a good ground. If voltage is absent or insufficient, check the circuit beginning with circuit protection devices. Repair as needed.

   Task Completed ___

10. If adequate voltage is present but the blower does not operate, test the blower motor by supplying voltage and ground directly to the motor. Repair as needed.

    Task Completed ___

11. Repeat steps 1-3 for the sleeper compartment if so equipped.

    Task Completed ___

Problems Encountered

_____
_____
_____

INSTRUCTOR EVALUATION

☐ 4 Mastered Task
☐ 3 Able to Perform Task Independently; Some Additional Training Suggested
☐ 2 Able to Perform Task with Close Supervision; Requires Additional Training
☐ 1 Unable to Perform Task
☐ 0 Not Attempted

Comments

_____
_____
_____

Instructor Name _____   Date _____

Instructor Signature _____

HEATING, VENTILATION, & AIR CONDITIONING
JOB SHEET #25

HVAC Electrical Control System Components

Name _____ Station _____ Date _____

NATEF Correlation

This Job Sheet addresses the following NATEF task(s):

D.1.3 Inspect and test A/C compressor clutch relays, modules, wiring, sensors, switches, diodes, and protection devices; determine needed action. (P-2)

Performance Objective(s)

Upon completion of this Job Sheet, you will be able to successfully diagnose the cause of failures in HVAC electrical and electronic control systems.

Tools and Materials
Truck with a functional HVAC system or equivalent mock-up
Appropriate service publications
Digital multimeter
Manifold gauge set or refrigerant recovery and recharging station
Electronic Service Tool (EST)
Hand tools

Protective Clothing/Equipment
Coveralls or shop coat
Safety glasses
Safety shoes
Rubber or nitrile gloves

PROCEDURE

1. Is this HVAC system computer controlled and networked to the data bus? If so, the first step in troubleshooting should be to connect an EST to the chassis data bus.

   Task Completed ___

2. Select MID 190 for the HVAC system.

   Task Completed ___

3. Scroll through the self diagnostic display and list any fault codes present.

   Task Completed ___

4. List the Failure Mode Indicator (FMI) for each code.

   Task Completed ___

5. What diagnostic tips and tests are recommended?

   _____
   _____

6. Using the service information provided for the truck you are diagnosing locate the wiring diagram for the air conditioning system.

   Task Completed ___

7. List the relays, modules, sensors, switches and diodes used in this system.

   _____
   _____
   _____
   _____

8. Locate these components on the truck.

   Task Completed ___

9. Using the multimeter provided determine whether the low pressure switch is closed. If it is open, check refrigerant pressure to determine whether there is sufficient refrigerant to close the switch. (If the truck uses either a binary switch or a trinary switch, use the wiring diagram to locate the correct terminals to test) If pressure is too low to close the switch correct the problem with the refrigerant system before proceeding.

   Task Completed ___

10. Remove the relay. Measure the resistance of the control circuit coil. Replace if needed.

    Task Completed ___

11. Replace the relay; using the trucks electrical circuit, energize the relay. Measure the voltage drop across the load carrying contacts. Replace if needed.

    Task Completed ___

12. Locate and remove the compressor clutch diode. Test and replace the diode if needed.

    Task Completed ___

13. Start the engine and select MAX A/C. Does the compressor clutch engage?

    _____

14. If the compressor clutch does not engage, use the wiring diagram and the diagnostic flow chart, trace the circuit and locate the source of the failure.

    Task Completed ___

15. With the system operating and the compressor clutch engaged, observe operation of the system. Does the electrical system control the A/C system normally? What additional repairs are needed?

    _____
    _____
    _____

Problems Encountered

_____
_____
_____

INSTRUCTOR EVALUATION

☐ 4 Mastered Task
☐ 3 Able to Perform Task Independently; Some Additional Training Suggested
☐ 2 Able to Perform Task with Close Supervision; Requires Additional Training
☐ 1 Unable to Perform Task
☐ 0 Not Attempted

Comments

_____
_____
_____

Instructor Name _____ Date _____

Instructor Signature_____

HEATING, VENTILATION AND AIR CONDITIONING
JOB SHEET #26

Air, Vacuum and Mechanical Controls

Name _____ Station _____ Date _____

NATEF Correlation

This Job Sheet addresses the following NATEF task(s):

D.2.1 Diagnose the cause of failures in HVAC air, vacuum, and mechanical switches and controls; determine needed action (P1)

Performance Objective(s)

Upon completion of this Job Sheet, you will be able to successfully examine a HVAC system with either air, vacuum, or mechanical switches and controls and locate the cause of failures. As well as, determine what repairs are needed to correct the failure.

Tools and Materials
Truck with a functional HVAC system or equivalent mock-up
Appropriate service publications
Digital multimeter
Electronic Service Tool (EST)
Vacuum gauge
Pressure gauge
Thermometer or temperature probe
Hand tools

Protective Clothing/Equipment
Coveralls or shop coat
Safety glasses
Safety shoes
Rubber or nitrile gloves

PROCEDURE

1. Using the service manual for this truck, describe the type of HVAC controls. List and describe the servos and actuating cables etc.. Include the sleeper controls, if equipped.

   _____
   _____

2. Locate and test the vacuum or pressure source. Describe the results of your test.

   _____

3. Turn on the ignition switch. Move the control through each of the available fan speeds. Which of the available speeds are present?

   _____

4. Listen to the sound of the fan motor. If the fan is noisy, perform any repairs needed.

   Task Completed ___

5. If one or more of the available fan speeds is absent, follow the diagnostic flow chart in the service manual to locate and correct the problem.

   Task Completed ___

6. If the fan does not operate, verify that adequate voltage is present at the blower motor and that it has a good ground. If voltage is absent or insufficient, check the circuit beginning with circuit protection devices. Repair as needed.

   Task Completed ___

7. If adequate voltage is present but the blower does not operate, test the blower motor by supplying voltage and ground directly to the motor. Repair as needed.

   Task Completed ___

8. When the blower is operating satisfactorily, move the control to direct air to each available outlet. Describe the amount of air produced at each outlet.

   _____
   _____

9. Start the engine and bring to operating temperature and set speed at about half throttle. Move the controls from maximum cooling through maximum heat. Describe the systems ability to control the temperature of the air produced at the outlets.

   _____
   _____

10. Use the diagnostic flow chart in the service manual to determine needed action. What is needed to correct the faults?

    _____
    _____

Problems Encountered

_____
_____
_____

INSTRUCTOR EVALUATION

☐ 4 Mastered Task
☐ 3 Able to Perform Task Independently; Some Additional Training Suggested
☐ 2 Able to Perform Task with Close Supervision; Requires Additional Training
☐ 1 Unable to Perform Task
☐ 0 Not Attempted

Comments

_____
_____
_____

Instructor Name _____ Date _____

Instructor Signature _____

HEATING, VENTILATION, & AIR CONDITIONING
JOB SHEET #27

Refrigerant Recovery, Recycling, and Handling

Name _____ Station _____ Date _____

NATEF Correlation

This Job Sheet addresses the following NATEF task(s):

- **E.1** Maintain and verify correct operation of certified equipment. (P-1)
- **E.2** Identify (by label application or use of refrigerant identifier) and recover A/C system refrigerant. (P-1)
- **E.3** Recycle refrigerant. (P-1)
- **E.4** Handle, label, and store refrigerant. (P-1)

Performance Objective(s)

Upon completion of this Job Sheet, you will be able to successfully maintain and use certified equipment to recover, and recycle the refrigerant.

Tools and Materials

Truck with a functional HVAC system or equivalent mock-up
Appropriate service publications
Copy of SAE standards J-1991, J-1989, and J-2211, J-2209, and J-1732
Equipment check-off list to verify compliance with SAE standard J-1991
A/C recovery/recycling station with instructions included
Refrigerant storage containers
Refrigerant identifier
Hand tools

Protective Clothing/Equipment

Coveralls or shop coat
Safety glasses
Safety shoes
Rubber or nitrile gloves

PROCEDURE

1. Refer to SAE Standard J-1991 provided and determine if the service equipment you are using complies in all respects. Fill out the check list provided.

   Task Completed ___

2. Are the truck's service port fittings correct for the type of refrigerant being used?

   _____

3. Identify the type of refrigerant in the system either through the use of a refrigerant identifier or by the under-hood label.

   Task Completed ___

4. Hook the service hoses up to the truck's service ports.

Task Completed ___

5. Following the instructions provided with the equipment you are using recover the refrigerant.

Task Completed ___

6. Follow the steps to recycle the refrigerant.

Task Completed ___

7. Describe how refrigerant is properly labeled and stored.

_____
_____
_____
_____

Problems Encountered

_____
_____
_____

INSTRUCTOR EVALUATION

☐ 4 Mastered Task
☐ 3 Able to Perform Task Independently; Some Additional Training Suggested
☐ 2 Able to Perform Task with Close Supervision; Requires Additional Training
☐ 1 Unable to Perform Task
☐ 0 Not Attempted

Comments

_____
_____
_____

Instructor Name _____ Date _____

Instructor Signature _____

HEATING, VENTILATION, & AIR CONDITIONING
JOB SHEET #28

Handle, Label, Store and Test Recycled Refrigerant for Non-condensable Gases

Name _____ Station _____ Date _____

NATEF Correlation

This Job Sheet addresses the following NATEF task(s):

E.5   Test recycled refrigerant for non-condensable gases. (P-1)

Performance Objective(s)

Upon completion of this Job Sheet, you will be able to successfully handle, label and store refrigerants and detect the presence of non-condensable gasses

Tools and Materials
Truck with a functional HVAC system or equivalent mock-up
Appropriate service publications
A/C recovery/recycling and charging station
Storage cylinders
Calibrated pressure gauges
Pressure/temperature chart
Temperature probe or thermometer
Hand tools

Protective Clothing/Equipment
Coveralls or shop coat
Safety glasses
Safety shoes
Rubber or nitrile gloves

PROCEDURE

1. Store the container of refrigerant to be tested in a location that is protected from the sun and has a stable temperature of 65 degrees F. or higher for at least 12 hours.

   Task Completed ___

2. Using a calibrated pressure gauge measure refrigerant pressure in the container. What is its pressure?

   _____

3. Measure the air temperature 4 inches away from the containers surface. What is its temperature?

   _____

4. Is the pressure of the refrigerant correct for pure refrigerant?

   _____

5. If the refrigerant pressure is higher than normal? What action is needed?

   _____

Problems Encountered

_____
_____
_____

INSTRUCTOR EVALUATION

☐ 4 Mastered Task
☐ 3 Able to Perform Task Independently; Some Additional Training Suggested
☐ 2 Able to Perform Task with Close Supervision; Requires Additional Training
☐ 1 Unable to Perform Task
☐ 0 Not Attempted

Comments

_____
_____
_____

Instructor Name _____ Date _____
Instructor Signature_____